COMPUTER SCIENCE, TECHNOLOGY AND APPLICATIONS

NEW STRATEGIES IN CLOUD COMPUTING

Computer Science, Technology and Applications

Additional books and e-books in this series can be found on Nova's website under the Series tab.

COMPUTER SCIENCE, TECHNOLOGY AND APPLICATIONS

NEW STRATEGIES IN CLOUD COMPUTING

LUC LARIVIÈRE
EDITOR

snova
New York

NOTICE TO THE READER

Library of Congress Cataloging-in-Publication Data

ISBN: 978-1-53616-907-2

Published by Nova Science Publishers, Inc. † New York

CONTENTS

Preface		**vii**
Chapter 1	Cloud Computing: Agencies Have Increased Usage and Realized Benefits, but Cost and Savings Data Need to Be Better Tracked *United States Government Accountability Office*	**1**
Chapter 2	From Cloud First to Cloud Smart *Suzette Kent*	**145**
Chapter 3	DoD Cloud Strategy *Department of Defense*	**167**
Chapter 4	The DOD's JEDI Cloud Program *Heidi M. Peters*	**193**
Chapter 5	Cross-Border Data Sharing under the CLOUD Act *Stephen P. Mulligan*	**199**
Chapter 6	Supreme Court to Hear Digital Privacy Case: Can the Government Obtain Emails Stored Abroad? *Austin D. Smith*	**237**

Index **243**

Related Nova Publications **251**

PREFACE

Cloud computing enables on-demand access to shared computing resources providing services more quickly and at a lower cost than having agencies maintain these resources themselves. Chapter 1 discusses selected agencies' progress in implementing cloud services, the extent to which those agencies increased cloud service spending and achieved savings or cost avoidances, and examples of agency-reported cloud investments with notable benefits.

The Office of Management and Budget (OMB) has developed a new strategy to accelerate agency adoption of cloud-based solutions: Cloud Smart. Chapter 2 reports on the strategy of successful cloud adoption: security, procurement, and workforce.

The Department of Defense (DoD) has entered the modern age of warfighting where the battlefield exists as much in the digital world as it does in the physical. Cloud is a fundamental component of the global infrastructure that will empower the warfighter with data and is critical to maintaining our military's technological advantage as reported in chapter 3. Chapter 4 discusses accelerated adoption of the Joint Enterprise Defense Infrastructure (JEDI) Cloud program.

In the Clarifying Lawful Overseas Use of Data (CLOUD) Act, Congress enacted one of the first major changes in years to U.S. law governing cross-border access to electronic communications held by private companies. Chapter 5 reports on the major components of the CLOUD Act. Chapter 6 discusses United States v. Microsoft Corp as it pertains to the CLOUD Act.

Chapter 1 - Cloud computing enables on-demand access to shared computing resources providing services more quickly and at a lower cost than having agencies maintain these resources themselves. In 2012, OMB began requiring agencies to assess all IT investments for cloud services. GAO was asked to review agencies' reported use of cloud services. This chapter discusses selected agencies' progress in implementing cloud services, the extent to which those agencies increased cloud service spending and achieved savings or cost avoidances, and examples of agency- reported cloud investments with notable benefits. GAO selected 16 agencies to review based on their fiscal year 2017 IT budgets and analyzed their use of cloud services, associated spending and savings data, and guidance for assessing investments for these services. GAO interviewed agency officials in charge of cloud services and reviewed pertinent documents to identify acquisitions with notable benefits. GAO also interviewed OMB staff about their agency's role in federal cloud computing and related OMB guidance.

Chapter 2 - In the Report to the President on Federal IT Modernization, released publicly in 2017 in accordance with Executive Order 13800, the Office of Management and Budget (OMB) pledged to update the Government's legacy Federal Cloud Computing Strategy ("Cloud First"). Fulfilling this promise, the Administration has developed a new strategy to accelerate agency adoption of cloud-based solutions: Cloud Smart. Developed nearly a decade after its predecessor, Cloud Smart equips agencies with actionable information and recommendations gleaned from some of the country's most impactful public and private sector use cases. Beyond Cloud First, which granted agencies broad authority to adopt cloud-based solutions, Cloud Smart offers practical implementation guidance for Government missions to fully actualize the promise and potential of cloud-based technologies while ensuring thoughtful execution that incorporates practical realities. The new strategy is founded on three key pillars of successful cloud adoption: security, procurement, and workforce. Collectively, these elements embody the interdisciplinary approach to IT modernization that the Federal enterprise needs in order to provide improved return on its investments, enhanced security, and higher quality services to the American people.

Chapter 3 - The Department of Defense (DoD) has entered the modern age ofwarfighting where the battlefield exists as much in the digital world as it does in the physical. Data and our ability to process data at the ready are differentiators to ensure mission success. Cloud is a fundamental component of the global infrastructure that will empower the warfighter with data and is critical to maintaining our military's technological advantage. The DoD Cloud Strategy reasserts our commitment to cloud and the need to view cloud initiatives from an enterprise perspective for more effective adoption. It recognizes our experience over the past five years and identifies seven strategic objectives along with guiding principles to set a path forward. It emphasizes mission and tactical edge needs along with the requirement to prepare for artificial intelligence while accounting for protection and efficiencies. The strategy drives implementation toward the enterprise cloud environment, an ecosystem composed of a General Purpose and Fit For Purpose clouds. It focuses implementation activities on two fundamental types of work: first is the stand up of cloud platforms ready to receive data and applications, and second is the ongoing work to migrate existing applications and to develop new applications in the cloud. This effort is a Department priority. As we execute this strategy, we will continue to seek the active participation and commitment of all DoD Components to realize the benefits of cloud as we operate on the 21st century battlefield.

Chapter 4 - In September 2017, Deputy Secretary of Defense (DSD) Patrick Shanahan issued a memorandum calling for the accelerated adoption of a Department of Defense (DOD)-wide cloud computing system. Under the Joint Enterprise Defense Infrastructure (JEDI) Cloud program, DOD seeks to "acquire a ... cloud services solution that can support Unclassified, Secret, and Top Secret requirements," with a focus on commercially available services. Significant industry and congressional attention has been focused on the JEDI Cloud contract.

Chapter 5 - Law enforcement officials in the United States and abroad increasingly seek access to electronic communications, such as emails and social media posts, stored on servers and in data centers in foreign countries. Because the architecture of the internet allows technology companies to store data at a great distance from the physical location of their customers, electronic communications that could serve as evidence of a crime often are not housed in the same country where the crime occurred. This disconnect has caused governments around the world, including the United States, to seek data stored outside their territorial jurisdictions. In the Clarifying Lawful Overseas Use of Data (CLOUD) Act, Congress enacted one of the first major changes in years to U.S. law governing cross-border access to electronic communications held by private companies. The CLOUD Act has two major components. The first facet addresses the U.S. government's ability to compel technology companies to disclose the contents of electronic communications stored on the companies' servers and data centers overseas. The Stored Communications Act (SCA) mandates that certain technology companies disclose the contents of electronic communications pursuant to warrants issued by U.S. courts based on probable cause that the communications contain evidence of a crime. But a dispute arose over whether warrants issued under the SCA could compel disclosure of data held outside the territorial jurisdiction of the United States. While the Supreme Court was set to resolve this issue in *United States v. Microsoft*, the CLOUD Act amended the SCA to require that technology companies provide data in their possession, custody, or control in response to an SCA warrant—regardless of whether the data is located in the United States. On April 17, 2018, the Supreme Court ruled that the change in law mooted the *Microsoft* case. The second facet of the CLOUD Act addresses the reciprocal issue of foreign governments' ability to access data in the United States as part of their investigation and prosecution of crimes. Prior to the CLOUD Act, foreign nations seeking data in the United States were required to request the assistance of the U.S. government through either mutual legal assistance treaties (MLATs) or judicial instruments known as letters rogatory. Requests under either instrument are reviewed by U.S. courts before disclosure to the foreign nation can be authorized, but U.S. and foreign officials criticized the processes as inefficient and unable to accommodate the increasing number of data requests in the digital era. The CLOUD Act responds to calls for modernization by authorizing the executive branch to conclude a new form of international agreement through which select foreign governments can seek data directly from U.S. technology companies without individualized review by the U.S. government. Agreements authorized by the CLOUD Act would remove legal restrictions on certain foreign nations' ability to seek data directly from U.S. providers in cases involving "serious crimes" when not targeting U.S. persons, provided the Executive has determined that the foreign nation's laws adequately protect privacy and civil liberties, among other requirements. While the CLOUD Act conditions approval of covered agreements upon a host of restrictions, commentators debate whether these agreements will provide adequate protections for privacy, human rights, and civil liberties. Law enforcement officials in the United States and abroad increasingly seek access to electronic communications, such as emails and social media posts, stored on servers and in data centers located in foreign countries. The architecture of the internet allows technology companies significant flexibility as to the geographic location where they may store collected data. As a result, electronic communications that may be evidence of a crime are not necessarily housed in the same country where the crime occurred. This disconnect has caused governments around the world, including the United States, to seek data stored outside their territorial jurisdictions in the course of law enforcement investigations. It also has led to debate over the extent to which national governments can compel private companies to disclose data stored in foreign nations and the degree to which civil liberties and privacy concerns should inform the proper procedure for sharing such data. In the United States, this debate largely has centered on the Stored Communications Act (SCA), which is part of the broader Electronic Communications Privacy Act (ECPA). Although the SCA generally prohibits certain technology companies from disclosing the contents of electronic communications to third parties, it mandates disclosure to the U.S. government pursuant to a warrant based on probable cause that the communications contain evidence of a crime. In *United States v. Microsoft Corp.*, the Supreme Court was set to address whether the United States could compel Microsoft to release emails housed in a data center in Ireland through a warrant issued under the SCA. But less than one month after oral argument, Congress passed and the President signed into law the Clarifying Lawful Overseas Use of Data Act (CLOUD Act) as part of the Consolidated Appropriations Act, 2018. The CLOUD Act amends the SCA and requires service providers subject to the SCA to release data in their possession, custody, or control in response to an SCA warrant—regardless of whether the data is located in the United States. After the U.S. government obtained a new warrant for the emails held in Ireland under the authority of the CLOUD Act, the Supreme Court deemed *Microsoft* moot. A second facet of the CLOUD Act addresses the reciprocal issue of foreign governments' desire to access data in the United States as part of their investigation and prosecution of crimes. Prior to the CLOUD Act, foreign nations seeking data in the United States generally were required to request the assistance of the U.S. government through either procedures established by mutual legal assistance treaties (MLATs) or judicial requests known as letters rogatory. Requests under either instrument are reviewed by U.S. courts before disclosure to the foreign nation is authorized, but U.S. and foreign officials have criticized these processes as inefficient and unable to accommodate the increasing cross-border data demands in the digital era. The CLOUD Act responds to calls for modernization by authorizing the executive branch to conclude a new form of international agreement through which select foreign governments can seek data directly from U.S. technology companies without undergoing individualized review by the U.S. government. Agreements authorized by the CLOUD Act would remove legal restrictions on certain foreign nations' ability to seek data directly from U.S. providers in cases involving "serious crimes" when not targeting U.S. persons, provided that the United States has determined that the foreign nation's laws adequately protect privacy and civil liberties, among other requirements. This chapter reviews the development of cross-border data sharing laws in criminal matters in the United States. It begins with an overview of ECPA and the SCA. Next, the report discusses the questions raised in the *Microsoft* litigation and the impact of the CLOUD Act on those issues. Finally, the report examines the new form of international agreements authorized by the CLOUD Act and the commentary on the benefits and drawbacks of the potential new international data sharing agreements.

Chapter 6 - Update: In United States v. Microsoft Corp., the Supreme Court was set to address whether a warrant issued under the Stored Communications Act (SCA) allowed U.S. law enforcement to compel Microsoft to hand over emails it stores on a server overseas. Following the publication of this Sidebar, and less than one month after the Court heard oral argument in Microsoft, Congress passed and the President signed into law the Clarifying Lawful Overseas Use of Data (CLOUD) Act as part of the Consolidated Appropriations Act, 2018. The CLOUD Act amended the SCA to require that technology companies provide data in their possession, custody, or control in response to an SCA warrant—regardless of whether the data is located in the United States. The government then obtained a new warrant pursuant to the CLOUD Act and the Supreme Court dismissed the case as moot. The original Sidebar post previewing the Microsoft case, published on February 23, 2018, is below. Can U.S. law enforcement use a warrant to compel Microsoft to hand over emails it stores on a server in Ireland? The Supreme Court may soon answer that question in a case, United States v. Microsoft, that could have major repercussions for both digital privacy rights and law enforcement interests. The case comes on appeal from a 2016 decision by the U.S. Court of Appeals for the Second Circuit (Second Circuit), discussed in a previous Legal Sidebar, which held that the government could not use the Stored Communications Act (SCA) to compel Microsoft to disclose emails it stored abroad. While both the government and Microsoft agree that the SCA does not apply overseas, the central disagreement before the Supreme Court is whether using the SCA to allow the government to access electronic communications held by a domestic entity but stored abroad would constitute a domestic or foreign application of the law. This Legal Sidebar provides background on the case, including an overview of the

SCA, discusses the specific issues presented by the Microsoft case, and explores the case's implications for Congress.

In: New Strategies in Cloud Computing ISBN: 978-1-53616-907-2
Editor: Luc Larivière

Chapter 1

CLOUD COMPUTING: AGENCIES HAVE INCREASED USAGE AND REALIZED BENEFITS, BUT COST AND SAVINGS DATA NEED TO BE BETTER TRACKED*

United States Government Accountability Office

ABBREVIATIONS

Agriculture	Department of Agriculture
CIO	chief information officer
Commerce	Department of Commerce
Defense	Department of Defense
DHS	Department of Homeland Security
Education	Department of Education
Energy	Department of Energy

* This is an edited, reformatted and augmented version of United States Government Accountability Office; Report to Congressional Requesters, Publication No. GAO-19-58, dated April 2019.

FedRAMP	Federal Risk and Authorization Management Program
FITARA	Federal Information Technology Acquisition Reform Act
GSA	General Services Administration
HHS	Department of Health and Human Services
IT	information technology
Justice	Department of Justice
Labor	Department of Labor
NIST	National Institute of Standards and Technology
OMB	Office of Management and Budget
SBA	Small Business Administration
SSA	Social Security Administration
State	Department of State
Transportation	Department of Transportation
Treasury	Department of the Treasury
VA	Department of Veterans Affairs

Why GAO Did This Study

Cloud computing enables on-demand access to shared computing resources providing services more quickly and at a lower cost than having agencies maintain these resources themselves. In 2012, OMB began requiring agencies to assess all IT investments for cloud services.

GAO was asked to review agencies' reported use of cloud services. This chapter discusses selected agencies' progress in implementing cloud services, the extent to which those agencies increased cloud service spending and achieved savings or cost avoidances, and examples of agency- reported cloud investments with notable benefits. GAO selected 16 agencies to review based on their fiscal year 2017 IT budgets and analyzed their use of cloud services, associated spending and savings data, and guidance for assessing investments for these services. GAO interviewed agency officials in charge of cloud services and reviewed pertinent documents to identify acquisitions with notable benefits. GAO also interviewed OMB staff about their agency's role in federal cloud computing and related OMB guidance.

WHAT GAO RECOMMENDS

GAO is making one recommendation to OMB on cloud savings reporting, and 34 recommendations to the 16 agencies on cloud assessments and savings. Fourteen agencies agreed with all recommendations, OMB and one agency neither agreed nor disagreed, and one (Defense) agreed with one recommendation but not the other. GAO continues to believe its recommendation to the department is appropriate.

WHAT GAO FOUND

The 16 agencies GAO reviewed made progress in implementing cloud computing services (cloud services)—namely, they established assessment guidance, performed assessments, and implemented these services—but the extent of their progress varied. To encourage cloud service acquisition, the Office of Management and Budget (OMB) began requiring agencies to assess all information technology (IT) investments for cloud services. However, only 10 of the 16 agencies reviewed had established assessment guidance. In addition, while the agencies assessed the majority of their planned fiscal year 2019 IT investments for cloud services, 12 agencies had not completed an assessment of 10 or more investments. Nevertheless, 10 of the agencies reported increasing their use of cloud services between fiscal years 2016 through 2019 (see figure). Six agencies noted that inconsistent reporting of cloud investments and investment consolidation impacted their reported percentage.

Further, the 16 agencies reported that they had increased their cloud service spending since 2015 and 13 of the 16 agencies had saved $291 million to date from these services. However, these agencies identified issues in tracking and reporting cloud spending and savings data, including not having consistent processes in place to do so. Agencies also noted that OMB guidance did not require them to explicitly report savings from cloud implementations and, therefore, they had to specifically collect this data to meet GAO's request. As a result of these identified issues, it is likely that agency-reported cloud spending and savings figures were underreported.

Officials from 15 of the 16 agencies reported that they had identified significant benefits from acquiring cloud services, including improved customer service and the acquisition of more cost-effective options for managing IT services. In addition, these agencies identified nine cloud investments that, among other things, enhanced the availability of weather-related information, facilitated collaboration and information sharing among federal, state, and local agencies related to homeland security, and provided benefits information to veterans, as examples of systems that realized these benefits. One agency reported that it had not realized benefits because it did not have any completed migration efforts.

April 4, 2019
Congressional Requesters

Over the past 2 decades, the federal government's increasing demand for information technology (IT) has led to a dramatic rise in operational costs to develop, implement, and maintain its computer systems and services. Furthermore, while federal agencies' IT use provides essential services affecting the health, economy, and defense of the nation, this use has also led to a reliance on custom IT systems that can be costly to maintain. Accordingly, as a result of these costs, the federal government invests approximately $90 billion annually in IT.

As part of a comprehensive effort to transform IT within the federal government, in 2010, the Office of Management and Budget (OMB) began requiring agencies to shift their IT services to a cloud computing option when feasible.[1] As defined by the National Institute of Standards and Technology (NIST), cloud computing is a means for enabling on- demand access to shared pools of configurable computing resources (e.g., networks, servers, storage applications, and services) that can be rapidly provisioned and released. This approach offers federal agencies a means to buy services more quickly and possibly at a lower cost than building, operating, and maintaining these computing resources themselves.

To encourage federal agencies to begin taking advantage of cloud computing services (cloud services), in February 2011, OMB issued a Cloud First policy that required each agency's chief information officer (CIO) to implement a cloud service whenever there was a secure, reliable, cost-effective option.[2] Since 2012, we have issued several reports on agencies' use of cloud services, as detailed later in this chapter.

You asked us to conduct a further review of agencies' reported use of cloud services. Specifically, our objectives were to: (1) evaluate selected agencies' progress in implementing cloud services, (2) review the extent to which selected agencies have increased spending on cloud services and achieved cost savings or avoidances, and (3) describe examples of cloud investments with significant or notable benefits that have been identified by selected agencies.

To address these objectives, we selected a sample of agencies based on the size of their total IT budget for fiscal year 2017. Specifically, we categorized each of the 24 *Chief Financial Officers Act* agencies[3] by the size of its IT budget: large (more than $3 billion), medium ($1 billion to $3 billion), and small (less than $1 billion), as reported on OMB's IT

[1] OMB, *25 Point Implementation Plan to Reform Federal Information Technology Management* (Washington, D.C.: Dec. 9, 2010).

[2] OMB, *Federal Cloud Computing Strategy* (Washington, D.C.: Feb. 8, 2011).

[3] The 24 major agencies listed in the *Chief Financial Officers Act of 1990* are the Departments of Agriculture, Commerce, Defense, Education, Energy, Health and Human Services, Homeland Security, Housing and Urban Development, the Interior, Justice, Labor, State, Transportation, the Treasury, and Veterans Affairs; the Environmental Protection Agency, General Services Administration, National Aeronautics and Space Administration, National Science Foundation, U.S. Nuclear Regulatory Commission, Office of Personnel Management, Small Business Administration, Social Security Administration, and U.S. Agency for International Development. 31 U.S.C. § 901(b).

Dashboard[4] for 2017. We then selected up to six agencies with the largest budgets from each budget category.

Using these criteria, we selected 16 agencies, including 7 agencies from our prior work on cloud services conducted in 2012 and 2014.[5] These agencies were the Department of Agriculture (Agriculture), Department of Commerce (Commerce), Department of Defense (Defense), Department of Education (Education), Department of Energy (Energy), Department of Health and Human Services (HHS), Department of Homeland Security (DHS), Department of Justice (Justice), Department of Labor (Labor), Department of State (State), Department of Transportation (Transportation), Department of the Treasury (Treasury), Department of Veterans Affairs (VA), General Services Administration (GSA), Small Business Administration (SBA), and Social Security Administration (SSA).

For our first objective, we obtained and analyzed IT Dashboard data related to the 16 selected agencies' use of cloud services for fiscal years 2016 through 2018 and projected use in 2019. We chose to begin with fiscal year 2016 because we had previously reported on federal agencies' use of cloud services through fiscal year 2014 and fiscal year 2015 data was not available. We reviewed agency responses that were submitted for fiscal years 2016 through 2019 as part of the annual budget submission process in order to determine whether a specific investment was using cloud services.

To ensure the accuracy and completeness of the selected agencies' data on the use of cloud services, we downloaded data from the IT Dashboard on October 3, 2017, March 7, 2018, and October 9, 2018. We took this step because agencies may update their data on a quarterly basis throughout the fiscal year. In addition, we presented the results of our analysis to officials in charge of cloud services within the Office of the CIO at each selected

---

[4] OMB's IT Dashboard is a public website that provides detailed information on IT investments at 26 federal agencies. See https://itdashboard.gov/.

[5] GAO, *Cloud Computing: Additional Opportunities and Savings Need to Be Pursued*, GAO-14-753 (Washington, D.C.: Sept. 25, 2014); and *Information Technology Reform: Progress Made but Future Cloud Computing Efforts Should be Better Planned,* GAO-12- 756 (Washington, D.C.: July 11, 2012).

agency. We asked these officials to verify the completeness and accuracy of this data and provide any updates as appropriate. Officials at all 16 agencies confirmed the total number of investments using cloud services for fiscal years 2016 through 2018 and their projected use for fiscal year 2019. Based on these steps, we determined that these data were sufficiently reliable to report on agencies' progress in using cloud services.

In addition, we compared each selected agency's cloud guidance to OMB's Cloud First guidance and interviewed officials in charge of cloud services within the Office of the CIO at each agency regarding their cloud data and guidance. We also interviewed OMB staff in the Office of E-Government and Information Technology regarding its guidance.

For our second objective, we obtained and analyzed IT Dashboard data related to the 16 agencies' spending on cloud services for fiscal years 2015 through 2018. We chose to begin with fiscal year 2015 because we had previously reported on federal agencies' spending on cloud services through fiscal year 2014. In order to determine actual cloud spending costs for each fiscal year, we used agency spending data reported each subsequent fiscal year (from fiscal years 2017 through 2018), as of October 5, 2018.

We also administered a data collection instrument to obtain and analyze spending and savings by the 16 selected agencies for fiscal years 2014 through 2018, as well as future planned costs. We requested that these agencies provide spending and savings data broken down by investment, as OMB only requires federal agencies to report total spending by cloud deployment model on the IT Dashboard, and agencies were not required to identify whether any reported savings were cloud-related.

We took steps to help ensure the reliability of the data we collected. First, to minimize errors that might occur from respondents interpreting our instrument differently from our intended purpose, we reviewed the data collection instrument with agency officials who would be completing the instrument during meetings in October and November 2017. Second, we reviewed the completed spreadsheets to identify missing data or other errors, and consulted with a GAO data quality expert about these issues as appropriate. We also reviewed the associated notes regarding agencies' qualifications of the provided data and followed up with agency officials to clarify the responses as appropriate.

Lastly, we presented the results of our analysis of IT Dashboard data and the data obtained from the data collection instrument to each of the selected agencies between June and August 2018. We asked the agencies to verify the completeness and accuracy of these data and provide any updates as appropriate. Each of the selected agencies provided updated information related to our data collection instrument, which we incorporated as appropriate. Based on the measures we took to ensure the reliability of the data reported by the agencies to us and on the IT Dashboard, we determined that the data were sufficiently reliable for the purpose of this chapter.

For the third objective, we obtained and reviewed available documentation discussing examples of cloud computing investments identified by the selected agencies as having produced significant or notable benefits and key practices that helped ensure the effort was successful, and interviewed officials from the Office of CIO or other components in charge of cloud services regarding these benefits. As part of this work, we asked officials within the Office of the CIO at each selected agency to identify up to three examples of investments that benefited from the acquisition of cloud services.

Fifteen of the sixteen agencies in our review identified at least one example, while one agency—HHS—reported that it did not have any examples of cloud investments that had produced significant or notable benefits because it did not have any completed migration efforts. Because of the open-ended nature of the 15 agencies' responses to our questions, we conducted a content analysis of the information we received in order to identify and summarize the benefits and key practices that were identified by the 15 agencies. In addition, to select systems or investments to profile, we reviewed the 34 examples provided by the 15 agencies and narrowed the list to 11 examples. In doing so, we sought to have a mix of systems that provided mission critical services to the agency or the public, illustrated a range of cloud computing benefits, and included detailed information on the benefits achieved from using cloud services.

In technical comments received on a draft of this chapter, two agencies provided new information regarding the use of cloud services for their systems that were profiled in appendix VI of the draft report. Based on the additional information provided by the two agencies, we determined there was no longer sufficient detail regarding what benefits were realized for these systems. Therefore, we removed the two agencies' profiled examples from the report in order to be consistent with our methodology for reporting examples of systems that had realized benefits from the acquisition of cloud services. We then notified both agencies of this decision. Further details on our objectives, scope, and methodology are included in appendix I.

We conducted this performance audit from September 2017 to April 2019 in accordance with generally accepted government auditing standards. Those standards require that we plan and perform the audit to obtain sufficient, appropriate evidence to provide a reasonable basis for our findings and conclusions based on our audit objectives. We believe that the evidence obtained provides a reasonable basis for our findings and conclusions based on our audit objectives.

BACKGROUND

Investments in federal IT have the potential to make agencies more efficient in fulfilling their missions by reducing costs and improving operational efficiencies. Each year, the federal government invests approximately $90 billion in IT, with about 75 percent reportedly spent on operating and maintaining existing systems. However, as we have previously testified, federal IT investments have too frequently failed or incurred cost overruns and schedule slippages while contributing little to mission-related outcomes.[6] As a result, the federal government has spent billions of dollars on failed and poorly performing IT investments. These

[6] GAO, *Information Technology: Further Implementation of Recommendations is Needed to Better Manage Acquisitions and Operations*, GAO-18-460T (Washington, D.C.: Mar. 14, 2018).

investments have often suffered from ineffective management of project planning, requirements definition, and program oversight and governance tasks.

Accordingly, in February 2015, we added improving the management of IT acquisitions and operations to our high-risk list—a list of agencies and program areas that have a higher potential for fraud, waste, abuse, and mismanagement, or are in need of transformation.[7] In introducing this high risk area, we specifically noted that agencies spend a significant portion of their budgets on the operations and maintenance of IT systems and need to effectively manage these investments in order to ensure they continue to meet agencies' needs and deliver value. We issued an update to our high-risk report in February 2017 and noted that, while progress has been made in addressing the IT acquisitions and operations high-risk area, significant work remains to be completed, including establishing action plans to modernize or replace obsolete investments.

In addition, over the last 3 decades, Congress has enacted several laws to assist agencies and the federal government in managing IT investments. For example, Congress enacted the *Clinger-Cohen Act of 1996*[8] to assist agencies in managing their investments. This act requires OMB to establish processes to analyze, track, and evaluate the risks and results of major capital investments in information systems made by federal agencies and report to Congress on the net program performance benefits achieved as a result of these investments.

Further, in December 2014, Congress enacted Federal Information Technology Acquisition Reform provisions (commonly referred to as FITARA) as a part of the *Carl Levin and Howard P. 'Buck' McKeon National Defense Authorization Act for Fiscal Year 2015*.[9] The act requires OMB, among other things, to develop standardized performance metrics, including for cost savings and cost avoidances, and to submit quarterly reports to Congress on cost savings and reductions in duplicative information technology investments.

---

[7] GAO, *High Risk Series: An Update*, GAO-15-290 (Washington, D.C.: Feb. 11, 2015).
[8] 40 U.S.C. § 11101, et. seq.
[9] Pub. L. No. 113-291, division A, title VIII, subtitle D, 128 Stat. 3292, 3438 (Dec. 19, 2014).

More recently, recognizing the challenges in modernizing government IT systems, in December 2017, Congress enacted the *Modernizing Government Technology Act as part of the National Defense Authorization Act for Fiscal Year 2018*.[10] This law authorizes all covered agencies[11] to establish an IT system modernization and working capital fund to, among other things, transition legacy systems to commercial cloud computing and other innovative commercial platforms and technologies using agency reprogrammed funds. The act also establishes a Technology Modernization Fund administered by the Administrator of General Services, in consultation with the CIO Council, which will provide funds to federal agencies for modernization efforts.

As of March 2019, the board that oversees the Technology Modernization Fund had awarded $60.87 million to four projects that plan to migrate or deploy systems to cloud services. Specifically,

- GSA's project, which received an award of $20.65 million, is intended to expedite the completion of a new software as a service solution for the agency's payroll and work schedule and leave management within 2 years.
- The Department of Housing and Urban Development's project, which received an award of $20 million, is expected to accelerate the migration of five of the agency's most critical business systems from an on-premise mainframe database to the cloud within the next 2 years.
- Energy's project, which received an award of $15.2 million, is intended to help the agency move 45 separate on-premise email systems to the cloud within the next 3 years.
- Agriculture's project, which received an award of $5 million, is intended to help the agency migrate 10 applications to a shared services cloud platform model.

---

[10] Modernizing Government Technology provisions of the *National Defense Authorization Act for Fiscal Year 2018*, div. A, title X, subtitle G (2017).

[11] The term covered agency refers to the 24 major agencies listed in the *Chief Financial Officers Act of 1990*. 31 U.S.C. § 901(b).

Overview of Cloud Services

One approach to improving the government's management of IT services is through cloud computing. As mentioned previously, cloud computing is a means for enabling on-demand access to shared pools of configurable computing resources (e.g., networks, servers, storage applications, and services) that can be rapidly provisioned. More specifically, purchasing IT services through a cloud service provider enables agencies to avoid paying for all the computing resources that would typically be needed to provide such services. This approach offers federal agencies a means to buy services more quickly and possibly at a lower cost than building, operating, and maintaining these computing resources themselves.

According to NIST, cloud computing offers federal agencies a number of benefits:

- On-demand self-service. Agencies can, as needed, provision computing capabilities, such as server time and network storage, from the service provider automatically and without human interaction.
- Broad network access. Agencies can access needed capabilities over the network through workstations, laptops, or other mobile devices.
- Resource pooling. Agencies can use pooled resources from the cloud provider, including storage, processing, memory, and network bandwidth.
- Rapid elasticity. Agencies can provision the resources that are allocated to match what actual resources are needed according to demand. This is done by scaling resources up or down by adding or removing processing or memory capacity, or both, according to demand.
- Measured service. Agencies can pay for services based on usage. This allows agencies to monitor, control, and generate reports,

providing greater transparency into the agency's use of cloud services.

As noted in NIST guidance, cloud service providers have established three types of service models that are offered to consumers:

- Infrastructure as a service. The service provider delivers and manages the basic computing infrastructure of servers, software, storage, and network equipment. The consumer provides the operating system, programming tools and services, and applications.
- Platform as a service. The service provider delivers and manages the infrastructure, operating system and programming tools and services, which the consumer can use to create applications.
- Software as a service. The service provider delivers one or more applications and all the resources (operating system and programming tools) and underlying infrastructure to run them for use on demand.

NIST has also defined four types of cloud deployment models, including:

- Private cloud. Service is set up specifically for one organization, although there may be multiple customers within that organization and the cloud may exist on or off the customer's premises.
- Community cloud. Service is set up for organizations with similar requirements. The cloud may be managed by the organizations or a third party and may exist on or off the organization's premises.
- Public cloud. Service is available to the general public and is owned and operated by the service provider.
- Hybrid cloud. Service is a composite of two or more of the three deployment models (private, community, or public) that are bound together by technology that enables data and application portability.

According to NIST guidance, these deployment models impact the number of consumers and the nature of other consumers' data that may be present in the cloud environment. A public cloud should not allow a consumer to know or control other consumers of a cloud service provider's environment. However, a private cloud can allow for ultimate control in selecting who has access to a cloud environment. Community clouds and hybrid clouds allow for a mixed degree of control and knowledge of other consumers. Additionally, the cost for cloud services typically increases as control over other consumers and knowledge of these consumers increase.

OMB and Past and Current Administrations Have Undertaken Efforts to Increase Use of Cloud Services

In December 2010, OMB made cloud computing an integral part of its *25 Point Implementation Plan to Reform Federal Information Technology Management.*[12] The plan called for the development of a government- wide strategy to hasten the adoption of cloud services. To accelerate the shift, OMB required agencies to identify three systems to migrate to cloud services, create a project plan for migration, and migrate all three systems by June 2012.

In February 2011, OMB issued the *Federal Cloud Computing Strategy,*[13] as called for in its 25-point plan. The strategy provided definitions of cloud services; benefits of cloud services, such as accelerating data center consolidations; a decision framework for migrating services to a cloud environment;[14] case studies to support agencies' migration to cloud services; and roles and responsibilities for federal agencies. For example, the strategy states that NIST's role is to lead and collaborate with federal, state, and local government agency CIOs, private

[12] OMB, *25 Point Implementation Plan to Reform Federal Information Technology Management.*
[13] OMB, *Federal Cloud Computing Strategy.*
[14] The decision framework, among other things, identifies several key areas for determining the readiness for moving to a cloud environment, including the ability of the cloud service provider to address government security requirements.

sector experts, and international bodies to identify standards and guidance and prioritize the adoption of cloud services.

Subsequently, in December 2011, OMB established the Federal Risk and Authorization Management Program (FedRAMP), a government-wide program to provide joint authorizations and continuous security monitoring services for cloud services for all federal agencies.[15] GSA initiated FedRAMP operations, which the agency referred to as initial operational capabilities, in June 2012.

In 2012, OMB began requiring agencies to evaluate each investment, or components or systems within the investment, for cloud services, regardless of the overall life-cycle stage of the investment.[16] Agencies were required to report the status of each investment's evaluation as part of the annual budget submission, as noted in OMB's annual capital planning guidance. Specifically, OMB required agencies to select an option regarding whether they had evaluated a cloud alternative and chosen a cloud alternative with a particular cloud deployment model or indicate that they had not yet evaluated the investment for cloud services. Starting in fiscal year 2018, OMB revised the options that agencies were to select from and required agencies to select an option regarding whether the investment, or a portion of the investment, was leveraging cloud computing, or indicate that cloud computing had not been considered for the investment.[17]

In 2012, OMB began requiring agencies to report associated cloud spending, as called for in its annual capital planning guidance.[18] For fiscal years 2015 through 2018, OMB's capital planning guidance required agencies to report their total cloud spending at the agency level based on the cloud deployment model, rather than by individual investment.[19]

---

[15] OMB, *Security Authorization of Information Systems in Cloud Computing Environments* (Washington, D.C.: Dec. 8, 2011). FedRAMP is governed by the Joint Authorization Board (composed of the CIOs from Defense, DHS, and GSA), the FedRAMP program management office, NIST, the Federal CIO Council, OMB, and DHS.

[16] OMB, *FY 2014 Guidance on Exhibits 53 and 300* (Washington, D.C.: Aug. 3, 2012).

[17] OMB, *FY 2018 IT Budget–Capital Planning Guidance* (Washington, D.C.: June 30, 2016).

[18] OMB, *FY 2014 Guidance on Exhibits 53 and 300.*

[19] OMB, *FY 2015 Guidance on Exhibits 53 and 300*; *FY 2016 IT Budget–Capital Planning Guidance* (Washington, D.C.: June 27, 2014); *FY 2017 IT Budget–Capital Planning*

Starting in fiscal year 2019, OMB will require agencies to report total cloud spending by investment and use the Technology Business Management Framework.[20] The Framework provides a cost taxonomy for agencies to use to manage the cost, quality, and value of their IT services. Specifically, agencies will be required to use a standard set of cost categories to group IT spending, including cloud-related spending. This new model is intended to increase the granularity in reporting of agency IT budget and spending data.

In addition, in May 2017, the administration established the American Technology Council to help transform and modernize federal IT and how the government uses and delivers digital services.[21] The President is the chairman of this council, and the Federal CIO and the United States Digital Service Administrator[22] are among its members.

Subsequently, in December 2017, the American Technology Council issued a *Report to the President on Federal IT Modernization* and made eight cloud computing-related recommendations that are relevant to the focus of our review.[23] For example, the report recommended that OMB issue two data calls to agencies in order to: (1) obtain a list of agency in-progress and pending projects for cloud migration; and (2) have agencies identify systems that have not yet migrated due to perceived or encountered difficulties. Based on the information provided, OMB would then assist agencies in making transition plans and work to remove obstacles in order to accelerate cloud adoption. In addition, the report recommended that OMB take action to update its guidance related to cloud computing and revise the Federal Cloud Computing Strategy that was previously issued in 2011.

Guidance (Washington, D.C.: June 16, 2015); and *FY 2018 IT Budget–Capital Planning Guidance*.

[20] *OMB, FY 2019 IT Budget–Capital Planning Guidance* (Washington, D.C.: Aug. 1, 2017).

[21] The American Technology Council was established under Executive Order on May 1, 2017 in order to promote the secure, efficient, and economical use of information technology within the federal government.

[22] The United States Digital Service is an office within OMB that aims to improve the most important public-facing federal digital services.

[23] White House American Technology Council, *Report to the President on Federal IT Modernization* (Washington, D.C.: Dec. 13, 2017).

According to staff in OMB's Office of E-Government and Information Technology, OMB has taken action to address these recommendations. For example, the staff reported that the two data calls were issued in December 2017 and staff are currently reviewing the information provided by agencies in response. In addition, OMB issued its draft strategy revision, the *2018 Federal Cloud Computing Strategy*, for comment on September 24, 2018.[24] This proposed Cloud Smart policy outlines a strategy for agencies to adopt cloud solutions that streamline transformation and embrace modern capabilities.

According to the draft strategy, Cloud Smart focuses on equipping agencies with the tools needed to make informative technology decisions in accordance with their mission needs. In addition, the draft strategy indicates that OMB intends to leverage private-sector solutions to provide the best services to the American people. The strategy also notes that the CIO Council and Chief Financial Officer Council are to work with OMB, GSA, DHS, and other federal entities to develop a work plan of actions and targeted policy updates that are to be delivered over the next 18 months. For more information about the current status of each of these eight cloud recommendations, as reported by OMB, please see appendix II.

Prior GAO Reports on Efforts to Implement Cloud Services

During the past several years, we reported on federal agencies' efforts to implement cloud services, and on the progress that oversight agencies have made to help federal agencies in those efforts. For example, in July 2012, we reported that the seven federal agencies we reviewed had made progress in meeting OMB's requirement to implement three cloud services by June 2012.[25] Specifically, the seven agencies had implemented 21 cloud services and spent a total of $307 million for cloud computing in fiscal year 2012—about 1 percent of their total IT budgets. In addition, while all

---

[24] OMB, *2018 Federal Cloud Computing Strategy* (draft), (Washington, D.C.: Sept. 24, 2018).

[25] GAO, *Information Technology Reform: Progress Made but Future Cloud Computing Efforts Should be Better Planned*, GAO-12-756 (Washington, D.C.: July 11, 2012).

seven agencies had submitted plans to OMB for implementing cloud solutions, all but one plan were missing key required elements. We made 14 recommendations to the seven agencies to develop planning information, such as estimated costs and legacy IT systems' retirement plans for existing and planned services. The agencies generally agreed with, and implemented, 13 out of 14 of our recommendations.

In September 2014, we reviewed the efforts of the same seven federal agencies again and found that each of them had implemented additional cloud services subsequent to our July 2012 report.[26] In particular, the total number of cloud services implemented by the seven agencies had increased by 80 services, from 21 to 101. The seven agencies' reported spending on cloud services had also increased by $222 million, from $307 million in 2012 to $529 million in 2014. However, this relatively small increase in cloud spending was attributed, in part, to the fact that these agencies had not considered cloud services for 67 percent of their investments. Accordingly, we recommended that the seven agencies assess their IT investments for suitability for cloud services. The agencies generally agreed with our recommendations and 6 of the agencies (Agriculture, DHS, GSA, HHS, SBA, and State) implemented all of our recommendations.

Further, in April 2016, we identified 10 key practices that federal and private-sector guidance noted should be included in service-level agreements in a contract when acquiring IT services though a cloud services provider.[27] However, our review of five agencies' (Defense, DHS, HHS, Treasury, and VA) cloud service contracts found that not all 10 key practices were included in these contracts. We therefore made recommendations to OMB to include all 10 key practices in future guidance to agencies. We also recommended that the five agencies incorporate these key practices as their contract and service level agreements expire. The agencies generally agreed with our

[26] GAO, *Cloud Computing: Additional Opportunities and Savings Need to Be Pursued*, GAO-14-753 (Washington, D.C.: Sept. 25, 2014).

[27] GAO, *Cloud Computing: Agencies Need to Incorporate Key Practices to Ensure Effective Performance*, GAO-16-325 (Washington, D.C.: Apr. 7, 2016).

recommendations and, to date, Defense and DHS have taken action to implement the recommendations.

More recently, in April 2017, we highlighted the results of a forum, convened by the Comptroller General on September 14, 2016, to explore challenges and opportunities for CIOs to improve federal IT acquisitions and operations—with the goal of better informing policymakers and government leadership.[28] Thirteen current and former federal agency CIOs, members of Congress, and private-sector IT executives who participated in the forum noted challenges with agency operations that could be addressed by migrating more services to the cloud. In their view, this approach would offer agencies a means to buy the services faster and possibly at a lower cost than through the traditional methods of building and maintaining systems.

In addition, forum participants noted the importance of federal agencies' IT procurement offices and processes evolving to align with new technologies, as agencies are not always set up to take advantage of cloud services. Lastly, forum participants said that, as the federal government is expected to increase its purchase of IT as a service with the move toward cloud computing, more oversight is needed to ensure that appropriate contracts are in place and appropriate oversight of performance occurs.

MOST OF THE SELECTED AGENCIES REPORTED MAKING PROGRESS IN IMPLEMENTING CLOUD SERVICES

The16 selected agencies reported making progress in implementing cloud services—namely, they established guidance for assessing investments for cloud services, performed those assessments, and implemented cloud services for their investments. However, the extent of these agencies' progress varied. Specifically, 10 of the 16 agencies established guidance for assessing all new and existing investments for

[28] GAO, *Information Technology: Opportunities for Improving Acquisitions and Operations*, GAO-17-251SP (Washington, D.C.: Apr. 11, 2017).

cloud services, while six agencies did not. In addition, while these agencies had assessed the majority of their investments for cloud services planned for fiscal year 2019, 12 agencies had not completed an assessment of 10 or more IT investments for cloud services. Lastly, 10 of the agencies reported a percentage increase in the use of cloud services from fiscal year 2016 through fiscal year 2019, while two agencies reported no percentage change and four agencies reported a decrease during this 4-year period.

About Two-thirds of the Selected Agencies Had Established Guidance for Assessing Investments for Cloud Services Suitability

OMB's Cloud First policy, issued in February 2011, requires each agency's CIO to implement a cloud service whenever there is a secure, reliable, cost-effective option to do so.[29] Further, subsequent OMB capital planning guidance, issued in 2014, requires agencies to evaluate each investment, or components or systems within the investment, for cloud services, regardless of the overall life-cycle stage of the investment.[30]

While OMB's guidance is not specific on how agencies should conduct these evaluations, GAO's Information Technology Investment Management framework[31] notes that organizations should have documented policies and procedures for the management oversight of IT investments, including the selection of investments and the evaluation of information technologies that have the potential to improve the organization's business.

Ten of the 16 agencies we reviewed had established guidance in accordance with OMB's requirement to assess new and existing IT investments for suitability for cloud services, as of August 2018. In

---

[29] OMB, *Federal Cloud Computing Strategy.*

[30] OMB, *FY 2016 IT Budget–Capital Planning Guidance.*

[31] GAO, *Information Technology Investment Management: A Framework for Assessing and Improving Process Maturity (Supersedes AIMD-10.1.23)*, GAO-04-394G (Washington, D.C.: March 2004).

particular, all 10 agencies' guidance required assessments of cloud service suitability for both new and existing IT investments.

However, the remaining six agencies did not have such comprehensive guidance in place. Rather, the guidance either required assessments of new or existing systems for cloud services but not both, or the guidance had not yet been established. Specifically, Labor's and SSA's guidance required assessments of new investments for cloud services but did not address assessments of existing investments. In addition, Energy's guidance required assessments of existing systems but not new acquisitions. Further, three agencies (Education, HHS, and Transportation) had not established guidance for assessing investments for cloud services.

The results of our analysis of agencies' guidance on assessing IT investments for cloud services are shown in Figure 1.

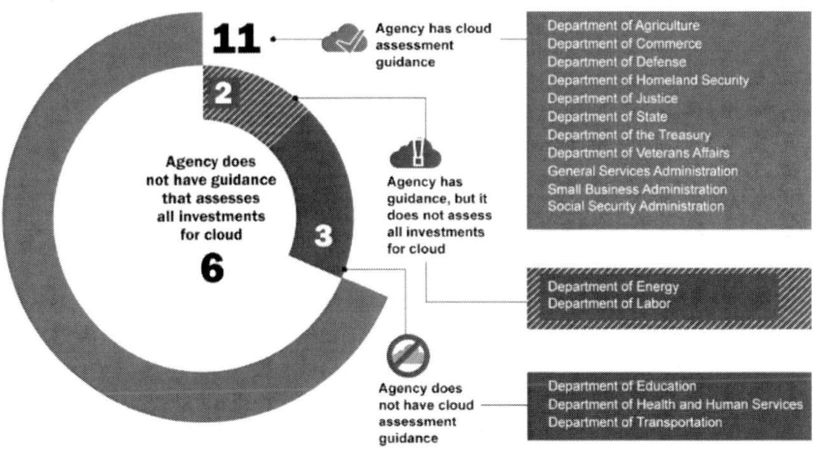

Source: GAO analysis of agency cloud guidance. | GAO-19-58.

Figure 1. Analysis of Selected Agencies' Cloud Guidance for Assessing Information Technology Investments for Cloud Computing Services, as of August 2018.

Agency officials in the Office of the CIO at the six agencies provided a variety of reasons for why they did not have guidance for assessing all investments for cloud services. Specifically, Labor officials reported that they had not included legacy applications in their guidance because not all applications should or could be migrated to the cloud. In addition, SSA officials reported that they were planning to assess all existing systems for cloud services, but had not determined a time frame for this review. Further, Energy officials reported that, while the agency was following OMB's Cloud First policy, it would need to establish guidance for assessing new investments for cloud services; however, a date for doing so had not been determined.

Transportation officials reported that they believed their guidance on managing cloud computing efforts was consistent with OMB's Cloud First policy and stated that they had no plans to develop additional guidance. However, our review of the agency's guidance found that it did not include any information regarding the assessment of investments for cloud services. Instead, the guidance only required that investments intending to use cloud services provide procurement and other cost information as part of the business case and use specified language in contracts with cloud service providers. Therefore, we believe that the guidance is not consistent with OMB's guidance requiring agencies to assess investments for cloud services.

Education officials reported that they were in the process of finalizing a policy and hoped to have it completed by the end of the year. In addition, HHS officials reported that they had explored developing some guidance regarding cloud services, but had not established any plans to do so.

As previously discussed, assessing all new and existing IT investments to determine whether they are suitable for cloud services is an important component of OMB's Cloud First policy. Until the six identified agencies update or establish guidance for assessing both new and existing investments for cloud services, they will not be positioned to ensure adequate implementation of OMB's Cloud First policy. Further, these agencies increase the risk that they will not be able to take advantage of cloud services to improve operational efficiencies and minimize costs.

Selected Agencies Assessed the Majority of Their IT Investments for Cloud Services

As noted previously, OMB's fiscal year 2016 IT capital planning guidance requires agencies to evaluate each investment for cloud services and report the status of this evaluation as part of the annual budget submission. Specifically, agencies were to respond to a question regarding whether they had selected cloud services for the investment, or components or systems within the investment, or, for example, report that the investment had not yet been assessed for cloud services. OMB publicly reports agencies' responses to this question on the IT Dashboard.

As of October 9, 2018, the 16 agencies in our review reported on the IT Dashboard that they had completed cloud assessments for 84 percent of their IT investments (5,180 out of a total of 6,157) planned for fiscal year 2019. Of these, two agencies (GSA and State) had completed an assessment of all investments. However, 12 agencies had not completed an assessment of 10 or more IT investments for cloud services.

Table 1 lists the number of IT investments at the 16 selected agencies for fiscal year 2019 that had been assessed for cloud services. The table also shows the number and percentage of investments that remained to be assessed.

Officials in the Office of the CIO at the 12 agencies provided a variety of reasons for why they had not assessed all investments for cloud services. For example, Agriculture officials reported that 21 of their 53 investments did not need assessments because the investments were not suitable for cloud services. The officials said they intended to update the IT Dashboard to reflect this change. Further, these officials stated that they planned to assess the remaining 32 investments by April 30, 2019.

Defense officials reported that the agency was in the process of adjusting its cloud strategy that was issued in January 2018 and intends to address investments that have not yet been evaluated. However, the officials stated that they had not established time frames for the evaluations. In addition, DHS officials reported that they were in the process of implementing their guidance and putting in place a new process for identifying planned acquisitions based on the phase in the acquisition life cycle. However, the officials had not identified a time frame for when the new process would be finalized or when all assessments of the investments would be completed.

Table 1. Status of Information Technology (IT) investment assessments for cloud computing services, as reported on the IT dashboard by 16 selected agencies for fiscal year 2019

	Number			Percentage
Agency	Total IT investments	Investments assessed for cloud services	Investments not assessed for cloud services	Investments not assessed for cloud services
Department of Agriculture	225	172	53	24
Department of Commerce	167	164	3	2
Department of Defense	2,735	2,498	237	9
Department of Education	146	96	50	34
Department of Energy	365	246	120	33
Department of Health and Human Services	727	614	113	16
Department of Homeland Security	391	339	52	13
Department of Justice	250	193	57	23
Department of Labor	158	146	12	8
Department of State	94	94	0	0
Department of the Treasury	273	126	147	54
Department of Transportation	357	288	69	19
Department of Veterans Affairs	62	36	26	42
General Services Administration	114	114	0	0
Small Business Administration	26	24	2	8
Social Security Administration	66	30	36	55
Total	6,157	5,180	977	16

Source: GAO analysis of IT Dashboard data as of October 9, 2018. | GAO-19-58.

Justice officials reported that, as they began the budget process, the agency planned to look at performing additional assessments of investments for cloud services. However, the officials provided no time frames for when these assessments would be completed. In addition, SSA officials stated that the agency planned to perform an assessment of current investments for cloud services. However, the officials reported that they had not established a time frame for completing these assessments.

Further, Treasury officials reported that, while the agency had established a process for assessing investments for cloud services, it did not set specific dates for when the assessments were to be conducted. These officials reported that they only conducted a cloud assessment if the agency determined that it would replace, redevelop, or retire an investment. However, Treasury's guidance is not consistent with OMB's requirement that agencies conduct an annual assessment of all investments, regardless of the overall life-cycle stage of the investment.

Many of the 16 agencies in our review have made progress in implementing cloud services by establishing guidance for assessing investments for cloud services and performing assessments. Even agencies that lacked formal guidance for performing an assessment have made progress in increasing the use of cloud services when the assessment was completed. Nevertheless, 12 agencies still need to assess a large number of their investments. Until these agencies assess their investments that have yet to be evaluated for cloud services, they may not know which investments are likely candidates for migration to cloud services. Moreover, these agencies will not be positioned to take advantage of operational efficiencies, cost savings, and other benefits from the use of cloud services.

Selected Agencies Have Increased Their Use of Cloud Services

As of October 9, 2018, the 16 agencies in our review reported on the IT Dashboard that 11 percent of their IT investments were projected to use cloud services for fiscal year 2019—an increase of 3 percentage points from fiscal year 2016 to fiscal year 2019. In addition, 13 out of the 16 agencies reported that they planned to increase their use of cloud services, in some cases, by as much as 20 percentage points or more, between fiscal years 2018 and 2019.

Table 2 lists the percentage of the selected agency IT investments that used cloud services for fiscal years 2016 through 2018 and are projected for 2019. (For additional details on the number of cloud investments and the total investments reported by each of the selected agencies for fiscal years 2016 through 2019, see appendix III.)

In addition, while the majority of agencies made progress in implementing cloud services between fiscal years 2016 and 2019, the extent of agencies' progress varied. Specifically, 10 of the 16 agencies reported an increase in the use of cloud services, with the percentage of increase varying from up to 10 percentage points to 20 or more percentage points.

For the remaining six agencies, two reported no change in the percentage of investments using cloud services and four reported a decrease in the overall percentage of cloud usage. Figure 2 shows the breakdown in the range of percentage point changes in the use of cloud services for agency investments for fiscal years 2016 through 2019, as reported on the IT Dashboard.

Officials in the Offices of the CIO, and Office of Information Technology, at the six agencies that reported no change or a decrease in their cloud investment percentages during this 4-year period provided a variety of reasons for why this was the case, or had no comments regarding the lack of change in their cloud investment percentages. Specifically, Energy officials reported that the agency had not shown an increase in the percentage of its cloud investments due to an IT portfolio optimization effort designed to consolidate the agency's cloud investments. According to the officials, this optimization effort was designed to reduce the total number of these investments during the 4-year period. As a result, this optimization effort affected the overall percentage of cloud investments. As for Labor, its officials did not offer any comments regarding the lack of an increase in cloud use during this period.

Table 2. Percentage of Information Technology (IT) investments that use cloud computing services, as reported on the IT dashboard by 16 selected agencies, for fiscal years 2016-2019 (projected)

Percent				
Agency	**Fiscal year 2016**	**Fiscal year 2017**	**Fiscal year 2018**	**Fiscal year 2019**
Department of Agriculture	10	11	20	24
Department of Commerce	15	15	23	27
Department of Defense	4	3	3	3
Department of Education	10	11	5	8
Department of Energy	9	11	7	9
Department of Health and Human Services	13	17	17	19
Department of Homeland Security	21	27	17	18
Department of Justice	9	10	9	14
Department of Labor	17	19	19	17
Department of State	12	16	8	18
Department of the Treasury	4	6	6	10
Department of Transportation	8	10	8	12
Department of Veterans Affairs	15	16	7	10
General Services Administration	17	23	35	38
Small Business Administration	14	15	19	15
Social Security Administration	14	15	13	35
Government percentage[a]	8	9	9	11

Source: GAO analysis of IT Dashboard data as of October 9, 2018. | GAO-19-58.

[a]The government percentage was calculated by dividing the total number of investments using cloud services from the total number of IT investments, as reported by the 16 agencies for each fiscal year. Based on the total number of investments that we reviewed, at least 44 percent of these investments were from Defense, but that agency's reported percentage of cloud investments (3 to 4 percent each fiscal year) was one of the lowest among the agencies in our review. As a result, Defense's reported percentage lowered the government percentage reported for each fiscal year.

In addition, Defense, DHS, and Education officials reported that staff in their agencies had inconsistently applied the definition of cloud computing, which had led to differences in identifying and reporting the number of cloud investments within their agencies during this period. Further, DHS officials noted that the ongoing addition, combination, completion, and cancellation of investments had contributed to the fluctuation in the number of cloud investments within their agency. Finally, VA officials reported that their cloud identification processes were maturing during this period and, as such, had resulted in different cloud investment counts.

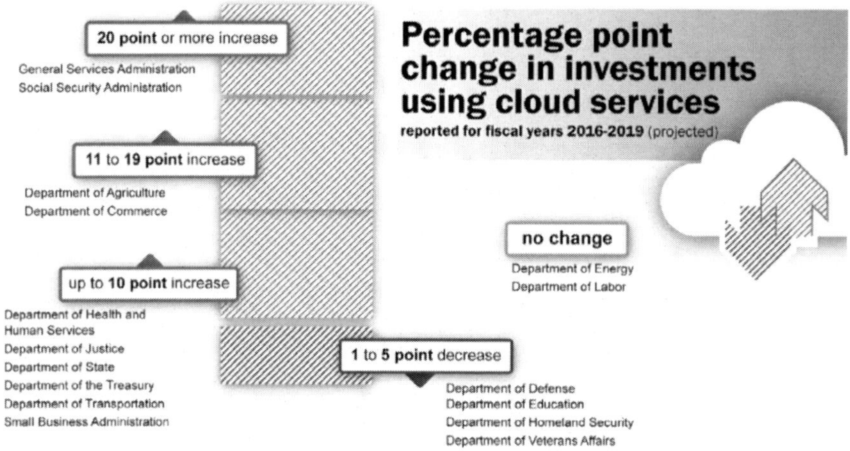

Source: GAO analysis of IT Dashboard data as of October 9, 2018. | GAO-19-58.

Figure 2. Information Technology (IT) Investments That Used Cloud Services, as Reported on the IT Dashboard by 16 Selected Agencies for Fiscal Years 2016-2019 (projected).

Some of the inconsistencies reported by agencies regarding the types of investments that they identified as being cloud investments may also be a result of OMB's changes to its guidance during this 4-year period. Specifically, OMB changed how agencies were required to report their use of cloud services in fiscal year 2018, and revised the options that agencies were to select from in order to identify and report the use of cloud services for each investment.

Going forward, several agencies reported that they intended to continue making progress in their implementation of cloud services beyond fiscal year 2019. For example:

- Education officials reported that the agency expected to increase cloud use significantly in 2019 and beyond due to an IT services contract award that is to support the migration of the agency's primary hosting infrastructure to the cloud.

- DHS officials reported that the agency had set aggressive goals for acquiring cloud services. Toward this end, the agency had initiated a cloud steering group and created a team with staff from all components. In addition, the officials reported that they planned to consolidate space in one data center and eliminate another data center, which would allow the agency to accelerate its migration to the cloud.
- Justice officials reported that they expected to increase spending on cloud services as the agency completed ongoing initiatives in 2019. In addition, the officials reported that they anticipated migrating the majority of the agency's unclassified data to the cloud in the next few years.
- VA officials reported that the agency planned to migrate at least 350 applications to the cloud by 2024.

Agencies' efforts to acquire additional cloud services and take advantage of improved efficiencies and cost savings should help to further improve their management of IT acquisitions and operations.

Agencies Have Increased Spending and Realized Savings from Using Cloud Services, but Spending and Savings Figures Are Underreported

The 16 agencies in our review made progress in implementing cloud services. Specifically, the 16 agencies reported that their spending on cloud investments had increased by over $1 billion between fiscal years 2015 and 2018 for investments with total life-cycle costs of $1 million or more. Nevertheless, the agencies reported that factors such as inconsistent tracking of spending data, along with confusion in interpreting OMB guidance, impacted the accuracy of their reported cloud spending data. In addition, 13 of the 16 agencies provided savings data indicating that they had saved hundreds of millions of dollars on cloud services, but agencies reported that they had problems with tracking this data. Further, six agencies reported that they had reinvested cloud savings into other IT modernization efforts or other improvements to IT services.

Selected Agencies Are Spending More on Cloud Services, but Do Not Have Complete Spending Data

OMB requires agencies to report spending on cloud services. Specifically, OMB's annual capital planning guidance for fiscal years 2015 through 2018[32] required agencies to report their total cloud spending on the IT Dashboard, although it did not require the information to be reported by investment.[33]

While the selected agencies' reporting on the IT Dashboard indicated that their percentage of total spending on cloud services generally remained constant during fiscal years 2015 through 2017. Specifically, the 16 agencies reported on the IT Dashboard that approximately 3 percent of their total IT spending each fiscal year during this period was spent on cloud services. For fiscal year 2018, agency-reported cloud spending through March 2018 was at 2 percent. Table 3 identifies the percentage of the selected agency spending on cloud services for fiscal years 2015 through 2018, as reported on the IT Dashboard. (For additional details on total agency cloud spending and total IT spending reported by each of the selected agencies for fiscal years 2015 through 2018, see appendix IV.)

However, the breakdown in spending by investment for cloud services with $1 million or more in life-cycle costs that the 16 agencies provided to us, showed that their spending on cloud investments had increased during fiscal years 2015 through 2018, and beyond (agencies generally submitted data on planned spending for one or more fiscal years beyond 2018).

---

32 OMB, *FY 2015 Guidance on Exhibits 53 and 300*; *FY 2016 IT Budget–Capital Planning Guidance; FY 2017 IT Budget–Capital Planning Guidance*; and *FY 2018 IT Budget–Capital Planning Guidance.*

33 As part of our review, we requested that agencies provide cloud spending data broken down by investment.

Table 3. Percentage of Total Information Technology (IT) cloud spending, as reported on the IT dashboard by the 16 selected agencies for fiscal years 2015-2018

Percent				
Agency	**Fiscal year 2015**	**Fiscal year 2016**	**Fiscal year 2017**	**Fiscal year 2018[a]**
Department of Agriculture	3	2	5	3
Department of Commerce	11	12	9	9
Department of Defense	4	4	1	1
Department of Education	9	9	10	1
Department of Energy	2	3	2	1
Department of Health and Human Services	0	1	4	3
Department of Homeland Security	4	4	2	2
Department of Justice	1	1	1	1
Department of Labor	8	9	6	4
Department of State	1	1	4	4
Department of the Treasury	4	4	3	1
Department of Transportation	1	1	1	1
Department of Veterans Affairs	1	1	7	6
General Services Administration	3	8	12	12
Small Business Administration	4	5	7	9
Social Security Administration	3	3	2	3
Government percentage[b]	3	3	3	2

Source: GAO analysis of IT Dashboard data as of October 5, 2018. | GAO-19-58.

[a]Although agency-reported cloud spending for fiscal year 2018 was obtained from the IT Dashboard on October 5, 2018, the agencies last updated their data between January through March 2018.

[b]The government percentage was calculated by dividing the total cloud spending from the total IT spending, as reported by the 16 agencies for each fiscal year.

Specifically, the agencies' provided data showed that total cloud spending for these investments was approximately $1.38 billion in fiscal year 2017—an increase of over $1 billion since fiscal year 2015. In addition, the 16 agencies' data indicated that they plan to spend over $3.2 billion on cloud services in fiscal year 2018 and beyond for these investments.

Table 4 summarizes the information provided to us on a breakdown of the 16 selected agencies' total spending for investments with $1 million or more in life-cycle costs for cloud services, from fiscal years 2015 through

2018 and beyond through fiscal year 2024, that was submitted to us. (For a list of the investments that have spent $1 million or more in life-cycle costs for cloud services, provided by each of the 16 selected agencies for fiscal year 2018, see appendix V.)

Table 4. Information on total spending for investments with $1 million or more in life-cycle costs for cloud services, provided by the 16 selected agencies for fiscal years 2015 through 2018 and beyond (in millions of dollars)

Agency	Fiscal year 2015	Fiscal year 2016	Fiscal year 2017	Fiscal year 2018 and beyond[a]
Department of Agriculture	84.91	102.96	242.23	272.70
Department of Commerce	2.53	21.75	33.74	92.65
Department of Defense	Not available	178.69	173.11	962.02
Department of Education	2.0	3.97	41.47	10.80
Department of Energy	10.80	13.42	17.06	28.37
Department of Health and Human Services	48.05	335.89	516.35	140.52
Department of Homeland Security	47.76	70.67	94.57	645.06
Department of Justice	25.99	40.38	36.08	39.61
Department of Labor	1.39	.37	1.90	5.81
Department of State	6.83	56.91	80.84	53.62
Department of the Treasury	20.27	24.62	32.24	93.82
Department of Transportation	15.31	25.02	26.98	30.46
Department of Veterans Affairs	3.74	3.47	15.57	782.49
General Services Administration	26.48	75.80	62.78	114.50
Small Business Administration	0.00	.07	3.85	9.31
Social Security Administration	0.00	0.00	3.01	17.73
Total	296.06	954.00	1,381.76	3,299.47

Source: GAO analysis of agency-provided cloud computing data. | GAO-19-58.

[a]In addition to providing planned spending on cloud services for fiscal year 2018, agencies generally submitted data on planned spending for one or more fiscal years beyond 2018 up through fiscal year 2024.

Officials in the Office of the CIO at all of the agencies in our review identified three factors that could affect the completeness of the cloud spending data provided to us and on the IT Dashboard: (1) spending data were not consistently tracked; (2) different methods were used to calculate cloud spending costs; and (3) interpreting changes in OMB and related guidance created confusion regarding what spending data should be tracked.

- Spending data were not consistently tracked. Defense officials reported, for example, that the agency had only begun tracking cloud spending in fiscal year 2016 and, therefore, spending data were not available for fiscal year 2015. In addition, VA officials reported that they were in the process of maturing their tracking of cloud spending data and, therefore, the agency did not have spending data available for the majority of their investments prior to fiscal year 2019. Further, Justice officials reported that the agency had been challenged to track cloud costs because the costs are based on fluctuating usage, rather than a flat rate.
- Different methods were used to calculate cloud spending costs. Some agencies reported that the data they provided to us included costs for items such as power usage and staff full-time equivalents, while other agencies told us that they only provided contract costs. In addition, some agency officials noted that they included in the provided spending figure, the additional costs for migrating the application to cloud services, while other officials said that these costs had not been included in their spending totals.
- Interpreting changes in OMB and related guidance created confusion regarding what spending data should be tracked. Agencies noted that OMB made changes to its guidance since 2015, including clarifications to the definition of cloud computing, changes to the definition and scope of cloud services and cloud spending, and changes to the guidance regarding what applicable costs should be included in spending totals—all of which created confusion regarding what investments and what costs should be tracked for cloud services. Defense officials also reported that the agency misinterpreted the NIST definition of cloud computing, and, as a result, Defense misreported that certain IT investments

were using cloud services, when these investments were not using these services. Defense officials reported that the agency had corrected this issue but it affected the total cloud spending reported during this period and led to the decrease in spending noted.

Based on our review of these factors reported by the 16 selected agencies, we identified issues with the completeness of the reported cloud spending data. Specifically, these factors increase the likelihood that all costs associated with spending on cloud services may have been incompletely captured by the 16 selected agencies in our review. As a result, agencies' reported total cloud spending on the IT Dashboard and the data provided to us is likely underreported.

Staff in OMB's Office of E-Government and Information Technology stated that agencies have previously reported challenges in breaking out cloud costs, particularly when the cloud acquisition is part of a larger contract. Given these challenges, the staff acknowledged that agency-reported cloud spending data are underreported and stated that the IT Dashboard reflects only a fraction of actual federal spending on cloud services. However, the staff stated that OMB's changes to its guidance, beginning in fiscal year 2019, should help to improve the reporting of cloud spending data. Specifically, beginning in fiscal year 2019, agencies will be required to report total cloud costs by investment, per OMB's IT capital planning guidance,[34] and use the Technology Business Management framework.[35]

Having complete data on spending for cloud services is critical in order to ensure that agencies can provide effective management and oversight of their cloud use, and that OMB and lawmakers can hold CIOs accountable for the performance of these cloud investments. The changes to OMB's guidance for fiscal year 2019 provide a key improvement for ensuring that agencies establish more consistent processes for reporting on cloud-

---

[34] *OMB, FY 2019 IT Budget–Capital Planning Guidance.*

[35] Under the Technology Business Management framework, agencies will be required to use a standard set of cost categories to group IT spending, including cloud-related spending. This new model is intended to increase the granularity in reporting of agency IT budget and spending data.

spending and should help agencies improve the completeness of the cloud-spending data that they report to OMB.

Selected Agencies Report Saving Approximately $291 Million from the Use of Cloud Services but Acknowledge Data Are Incomplete

Since 2013, OMB has required agencies to report quarterly on their total savings and cost avoidances[36] from implementing OMB's IT reform initiatives, including savings realized from the migration to cloud services.[37] Specifically, agencies are required to report actual and planned savings from implementing these initiatives in a quarterly submission and identify which implementation of an OMB initiative resulted in the reported savings. Despite this, in the reporting mechanism, agencies can only associate specific savings with certain OMB initiatives, a list that does not include the migration to cloud services.[38] *Standards for Internal Control in the Federal Government* emphasizes that management should track the actual performance of key initiatives in order to ensure that these activities are meeting plans, goals, and objectives, and in doing so, management should use quality information.[39]

---

[36] As defined in Circular A-131, OMB defines cost savings as a reduction in actual expenditures below the projected level of costs to achieve a specific objective. Cost avoidances are defined as results from an action taken in the immediate time frame that will decrease costs in the future. See OMB, *Value Engineering*, Circular A-131 (Revised) (Washington, D.C.: Dec. 26, 2013).

[37] OMB, *Fiscal Year 2013 PortfolioStat Guidance: Strengthening Federal IT Portfolio Management*, Memorandum M-13-09 (Washington, D.C.: Mar. 27, 2013).

[38] OMB requires agencies to report quarterly integrated data collection submissions using a standardized text format that allows agencies to submit data electronically. OMB provides guidance on submitting reported savings data through the CIO Council's website at https://management.cio.gov/schema/#savings. The available options for agencies to select in reporting savings include data center, digital services, commodity IT, PortfolioStat, software license management, or other.

[39] GAO, *Standards for Internal Control in the Federal Government*, GAO-14-704G (Washington, D.C.: September 2014).

Table 5. Information on total savings and cost avoidances for investments with $1 million or more in life-cycle costs for cloud services, provided by the 16 selected agencies for fiscal years 2014 through 2018 and beyond (in millions of dollars)

Number			Dollars (millions)		
Agency	**Current cloud investments**	**Current cloud investments with reported savings**	**Fiscal years 2014-2016**	**Fiscal year 2017**	**Fiscal years 2018 and beyond[a]**
Department of Agriculture	57	11	4.21	$7.11	18.74
Department of Commerce	16	2	22.98	15.20	4.83
Department of Defense	115	0	0.00	0.00	0.00
Department of Education	5	2	2.23	1.61	1.12
Department of Energy	40	16	33.82	6.52	9.04
Department of Health and Human Services	61	17	36.84	17.89	57.80
Department of Homeland Security	51	3	8.98	8.06	1.66
Department of Justice	29	1	3.47	4.70	4.88
Department of Labor	3	3	3.78	1.92	6.28
Department of State	16	0	0.00	0.00	0.00
Department of the Treasury	17	7	2.04	2.46	16.40
Department of Transportation	13	2	4.86	2.20	0.02
Department of Veterans Affairs	42	2	0.10	0.00	2.10
General Services Administration	16	9	79.53	20.89	26.12
Small Business Administration	4	2	0.00	0.00	1.35
Social Security Administration	3	0	0.00	0.00	0.00
Total	488	77	202.83	88.58	150.30

Source: GAO analysis of agency-provided cloud computing data. | GAO-19-58.

[a]In addition to providing information on planned savings or cost avoidances on cloud services for fiscal year 2018, agencies generally submitted data on planned savings for one or more fiscal years beyond 2018.

Thirteen of the 16 agencies in our review provided savings data to us for at least one cloud investment with life-cycle costs of $1 million or more for cloud services during fiscal years 2014 through 2018.[40] In total, the 13 agencies' provided data showed that they had accrued approximately $291 million in savings or cost avoidances using cloud services since 2014. In addition, the agencies' data indicated that they planned to save at least $150 million in fiscal year 2018 and beyond (agencies generally submitted data on planned savings for one or more fiscal years beyond 2018).

However, agency officials from the 13 agencies stated that, while they were able to provide some savings data, these data are only tracked on an ad hoc basis for certain cloud investments. In addition, officials from three agencies (Defense, State, and SSA) stated that they could not provide savings data for any of their cloud investments. As a result, the 16 agencies were unable to provide savings or avoidance data for 411 out of 488 investments (84 percent) that we reviewed.

Table 5 shows, for the selected agencies in our review, the breakdown in total agency savings and cost avoidances for fiscal years 2014 through 2018 and beyond for investments with $1 million or more in life-cycle costs for cloud services.

Officials in the Office of the CIO at the 16 agencies identified three factors that impacted their efforts to provide data on savings or cost avoidances for cloud computing investments: (1) savings data were not systematically tracked or were hard to track; (2) deploying or migrating systems to the cloud had resulted in no cost savings; and (3) OMB does not require agencies to identify savings associated with cloud services as part of reported savings.

- Savings data were not systematically tracked or were hard to track. Defense, Treasury, and VA officials reported that their investment management systems did not have the capability to track cloud savings or avoidance data. In addition, GSA officials reported that, while their system had the capability to track cost savings data, the

[40] As part of our work, we requested agencies provide cloud savings data broken down by investment.

agency did not capture and track realized cloud savings in a consistent format. GSA officials stated that they were in the process of implementing the Technology Business Management Framework, which they expected would improve the collection of these data. However, the officials did not identify a time frame for when this framework was to be implemented.

- Education officials reported that the agency did not provide cost savings data for those investments where cost savings targets had not been established or anticipated. SBA officials reported that investments with two cloud providers had only been recently made so the agency could not yet make a reasonable assessment of savings. Further, Agriculture officials reported that the agency had a hard time tracking the savings from certain investments because the process for formulating the overall agency budget was different than the process for determining savings from cloud implementation.

State officials reported that they were in the process of developing the capability to collect and track savings data from using cloud services but did not have any reliable data to provide during our review. In addition, Energy officials reported that their agency intended to establish review processes in the coming year to ensure that costs, cost savings, and cost avoidances were tracked for all cloud investments. As part of this process, the agency intended to work closely with its components to ensure that there was a consistent application of definitions for cloud spending and savings. However, the officials did not identify a specific time frame for when the agency expected the new process to be completed.

Lastly, HHS officials reported that the agency did not expect to track cost savings beyond the FITARA requirements. FITARA requires the reporting of savings associated with two OMB initiatives—data center consolidation and PortfolioStat. However, per M-13-09, OMB requires agencies to report savings associated with all of its initiatives.[41] As such, HHS's tracking of savings is not consistent with OMB's guidance.

- Deploying or migrating systems to the cloud resulted in no cost savings. Treasury officials reported that their agency had not realized any cost savings from the migration of certain investments because the acquisition of cloud services either had allowed the agency to purchase additional capabilities that the previous system did not have, or the agency had continued to operate the previous system at the same time as the new cloud system for a period of time. In addition, Commerce officials reported that their agency had not realized any cost savings for some investments because acquiring cloud services required that new business and performance requirements be put in place, which resulted in no overall savings for these investments. Further, DHS and SSA officials reported that a number of their investments were new applications that were developed and deployed in the cloud. As such, there were no costs from a prior system that could be compared with the costs to maintain the new system using cloud services; thus, there were no associated cost savings or avoidances.
- OMB does not require agencies to identify savings associated with cloud services as part of reported savings. Officials from Agriculture, Justice and Transportation noted that, while OMB requires agencies to report savings, current reporting instructions do not specifically require the identification and reporting of cloud savings as a separate category of cost savings and avoidance. In this regard, OMB's guidance requires agencies to identify which OMB initiative resulted in the reported savings, but the available options for agencies to choose from do not include cloud services.[42] Accordingly, officials from these agencies stated that

---

[41] OMB, Memorandum M-13-09.

[42] OMB requires agencies to report quarterly integrated data collection submissions using a standardized text format that allows agencies to submit data electronically. OMB provides guidance on submitting reported savings data through the CIO Council's website at https://management.cio.gov/schema/#savings. The available options for agencies to select in

> they either reached out to their components to try and collect this information or had to review their investments to determine whether there were any cloud savings, to be able to provide this information to us.

Based on our review of the factors that impacted the selected agencies' efforts to provide savings data, we identified issues with the completeness of the savings data. Specifically, challenges identified by the selected agencies in systematically tracking savings data, and the lack of a specific OMB requirement to report savings associated with cloud services, increase the likelihood that all savings associated with cloud services may have been incompletely captured by the agencies that provided these data. As a result, agencies' reported cloud savings data on the IT Dashboard and the data provided to us is likely underreported.

Staff in the Office of E-Government and Information Technology stated that, while agencies are required to report total savings related to OMB initiatives, the format is left to agency discretion. In addition, OMB staff confirmed that agencies do not have to specifically identify savings related to cloud computing unless they choose to do so. OMB staff further said that they do not require a specific format for reporting savings in order to minimize the burden on agencies in reporting this information.

While OMB's effort to minimize the reporting burden on agencies is appropriate, the lack of an explicit requirement to identify reported savings associated with cloud services has contributed to agencies not consistently tracking these savings. In addition, while OMB has taken steps to ensure more accuracy and granularity in agency reporting of cloud investment spending data in fiscal year 2019, there has not been a corresponding effort to ensure better reporting of cloud savings data. As a result, OMB and Congress may not have sufficient data to see the results of key initiatives, like Cloud Smart, and understand whether agencies are achieving savings using cloud services.

---

reporting savings include data center, digital services, commodity IT, PortfolioStat, software license management, or other.

Since 2013, OMB has required agencies to report on the savings resulting from implementation of its key IT reform initiatives. Although OMB does not provide the means for agencies to explicitly identify cloud-related savings, it is nevertheless important for agencies to take steps to fully track savings and cost avoidances from cloud computing acquisitions in order to ensure effective oversight and management of these initiatives. However, until OMB establishes a specific cloud savings reporting requirement, and until these agencies establish a consistent and repeatable mechanism to track these savings and cost avoidances, the agencies may lack sufficient information on the results of cloud acquisitions to date and the data necessary to make decisions regarding future cloud acquisitions.

Agencies Reported Reinvesting Cloud Implementation Savings into IT Modernization or Other Improvement Efforts

In 2017, Congress enacted what is known as the *Modernizing Government Technology Act*,[43] which authorized covered agencies[44] to establish an IT system modernization and working capital fund. This fund was to be used to, among other things, transition legacy IT systems to commercial cloud computing and other innovative commercial platforms and technologies using agency reprogrammed funds[45] or amounts made available to the IT working capital fund through discretionary appropriations.

Regardless of the extent of agencies' processes for tracking savings obtained from using cloud services, officials in the Office of the CIO at six agencies in our review (Education, GSA, Labor, SBA, SSA, and Treasury) reported that they have reinvested these savings into other IT modernization efforts or other improvements to IT services. For example:

[43] Modernizing Government Technology provisions of the *National Defense Authorization Act for Fiscal Year 2018*, Pub. L. No. 115-91, div. A., title X, subtitle G, § 1076, 131 Stat. 1283, 1586 (2017).

[44] The term ''covered agency'' refers to each agency listed in 31 U.S.C. § 901(b).

[45] Reprogramming is defined as shifting funds within an appropriation or fund account to use them for purposes other than those contemplated at the time of appropriation.

- Education officials reported that $498,000 in fiscal year 2018 cloud savings was used to modernize the agency's network infrastructure in order to provide increased multipath bandwidth and software that automatically routes traffic if network issues occur.
- GSA officials reported that the agency used the savings from replacing the agency's legacy on-premises email program with a cloud-based email system to implement a modern enterprise collaboration platform, email, and document storage system. According to the officials, the move to the cloud helped improve the agency's flexibility (the new system is accessible from any device, at any time, and from any location), productivity, and cost-effectiveness. As part of this effort, the officials reported that the savings were managed using the agency's working capital fund.
- Labor officials reported that their agency is using savings and cost avoidances to partially fund an initiative to consolidate cloud services within the agency in order to provide future secure cloud services and establish an enterprise contract vehicle to obtain cloud services. The officials noted that this investment is intended to allow the agency's components to leverage a cloud authority to operate, obtain competitive pricing for, and establish communications to cloud service providers. In addition, Labor officials reported that the agency has established a working capital fund that is to be used to manage the savings from cloud and shared services.
- SBA's CIO reported that the agency reinvested $7.8 million in savings from efforts to consolidate data centers to the cloud toward the implementation of other enterprise-wide modernization efforts. In addition, the agency had used the savings for the deployment and migration of additional applications to cloud services. Specifically, the CIO reported that the savings were used to design and architect cloud services, roll out the agency's update of key operating system and office applications, decommission obsolete data center assets, reduce overlapping technologies, and enhance

security and compliance capabilities with new enterprise tools and network monitoring. In addition, the CIO stated that, by using the savings from the data center consolidation, SBA has been able to undertake all of the agency's cloud modernization efforts with no additional budgeted funding.

We have previously reported[46] that significant work remains to ensure that agencies improve their management of IT acquisitions and operations, including modernizing or replacing obsolete IT investments. It is encouraging that several agencies have reinvested savings from cloud initiatives into other IT modernization efforts and, in some cases, have taken advantage of working capital funds authorized by Congress to do so. Having complete information on the savings or avoidances that result from cloud initiatives and using those savings to further IT modernization efforts is critical to ensuring the transformation of IT services across the federal government in the future.

SELECTED AGENCIES HAVE REALIZED BENEFITS FROM CLOUD SERVICES

Officials from 15 of the 16 agencies in our review reported that they had realized several significant benefits from the adoption of cloud services, ranging from improvements in the delivery of IT services to increasing the efficiency of operations and systems.[47] In addition, the 15 agencies noted that certain key practices enabled them to realize these benefits through the successful implementation of cloud services. These practices included establishing new governance planning activities and policies, reorganizing the management of agency IT resources, and having executive leadership involved to help drive acquisition efforts.

---

[46] GAO-17-317 and GAO-15-290.

[47] One agency in our review reported that it had not realized any significant benefits from acquiring cloud services.

Cloud Services Aid with IT Efficiency, Cost Savings, and System Modernization

Officials in charge of cloud services at 15 of the 16 agencies in our review reported that they had identified five significant or notable benefits as a result of acquiring cloud services.[48] Specifically, the 13 agencies reported that they had improved customer experiences through better design and performance of business systems and customer websites.

Table 6. Significant or notable benefits reported by 15 selected federal agencies from the acquisition of cloud computing services

Benefit identified	Number of agencies identifying benefit
Improved the delivery of IT services through more flexible and scalable IT resources	15
Helped reduce the cost of provisioning IT infrastructure and managing IT services	15
Increased the efficiency of operations and systems through system performance improvements and automated functions	15
Enhanced the customer's experience by improving system design and usability	13
Strengthened mission assurance through streamlined security resources and improved backup capabilities	13

Source: GAO analysis of agency data. | GAO-19-58.

Note: Some agencies reported multiple benefits from the acquisition of cloud computing services. The 15 agencies include the Departments of Agriculture, Commerce, Defense, Energy, Education, Homeland Security, Justice, Labor, State, Transportation, the Treasury, and Veterans Affairs; and the General Services Administration, Small Business Administration, and Social Security Administration.

In addition, all 15 agencies reported that they were able to procure more flexible and scalable IT resources,[49] and reduce the cost of

[48] The 15 agencies were Agriculture, Commerce, Defense, DHS, Energy, Education, GSA, Justice, Labor, SBA, SSA, State, Transportation, Treasury, and VA. HHS officials reported that they had no examples of cloud migrations or deployments that produced benefits because it did not have any completed migration efforts.

[49] Cloud computing allows an organization the ability to scale resources up or down by adding or removing processing or memory capacity, or both, according to demand. This enables the

provisioning infrastructure and managing services. Table 6 lists the five significant or notable benefits reported by the 15 agencies and the number of agencies that reported each benefit. The discussion that follows the table provides examples of each of the five agency-reported benefits from the acquisition of cloud services.

Procuring Cloud Services Improved the Delivery of IT Services

Officials in the Office of the CIO at the 15 agencies reported that acquiring cloud services had allowed them to procure IT resources that were more flexible and scalable than the prior legacy infrastructure. For example, officials in Labor's Office of the CIO reported that they had acquired cloud services to address seasonal demands for system processing. By eliminating the need to purchase additional servers and other equipment that would go unused during the rest of the year, Labor officials reported that cloud services allow the agency to scale resources up during these periods of increased processing and then scale the resources back down when the excess capacity is no longer needed.

In addition, officials in DHS's Office of the CIO reported that, in 2012, they had acquired software as a service for the agency's virtual desktop solution. This new service provided six agency components access to virtual secure desktop operating systems and applications. By eliminating the need for users to be physically present in a specified location in order to perform work activities, DHS officials reported that cloud services had improved the ability to quickly respond to the agency's mission needs and provided teleworking capabilities. In addition, the officials reported that the solution streamlined the process of provisioning network access between agency components and other external agencies.

Acquiring Cloud Services Helped Agencies Reduce the Cost of IT Services

Officials in the Office of the CIO at the 15 agencies reported that acquiring cloud services had allowed them to procure more cost-effective

---

organization to provision the resources that are allocated to match what actual resources are needed.

options for provisioning IT infrastructure and managing IT services. For example, officials in Education's Office of the CIO reported that, by migrating the Institute of Education Sciences'[50] data center to the cloud in 2014, the agency had saved approximately $3.3 million in cost avoidances annually for the last 3 years from not having to pay prior data center hosting charges. In addition, Education officials reported that the agency had saved $11.6 million between fiscal years 2013 and 2018 by eliminating contractor website hosting.

In addition, officials in Energy's Office of the CIO reported that the agency saved $900,000 in fiscal years 2013 to 2014 by transitioning to a cloud-based platform for managing IT services, such as asset management. Acquiring the software and platform as a service reduced or eliminated the costs of administering the agency's on-premise legacy infrastructure and associated software licensing fees.

Using Cloud Services Increased the Efficiency of Agency Operations and Systems

Officials in the 15 agencies' Offices of the CIO reported that acquiring cloud services had allowed them to streamline or improve systems and automate business processes and other functions. For example, officials in State's Office of the CIO reported that the agency had previously relied on paper-based and manual processes for completing employee requests for, among other things, leave, training, personal identification cards, and other general services. By acquiring software and platform as a service, State implemented an electronic application that replaced over 800 paper forms used to make these requests, without the time and cost of developing an application themselves. As a result, the officials reported that they estimate the application has saved more than 50,000 hours of staff time since its deployment by streamlining the request process, automatically populating common data fields, and improving support options.

---

[50] The Institute of Education Sciences provides education data and research that addresses school readiness and education from infancy through adulthood. The Institute's work includes, among other things, conducting surveys, sponsoring research projects, funding development and testing of approaches for improving education outcomes, and providing data on how well the United States is educating its students.

In addition, officials in Treasury's Community Development Financial Institutions Fund[51] reported that their office acquired software as a service, which will enable them to reduce the number of legacy systems related to awards management from 17 to 2. These legacy systems had required staff to enter the same data in different systems and manually complete certification work tasks. By automating many of the manual review and compliance processes, the officials reported that the office saved approximately 650 staff hours in 2017.

Cloud Services Helped Agencies Enhance Their Customer Service

Officials in the Office of the CIO at 13 agencies reported that acquiring cloud services had allowed their agencies to improve system design and usability, which helped to enhance their customer service. For example, VA officials reported that they had deployed a website in the cloud, Access to Care,[52] which included detailed data on the wait times and quality-care metrics at local hospitals. Doing so enabled veterans to be able to make better decisions about their health care options. By acquiring cloud services, VA officials reported that they had developed and deployed the new Access to Care website in approximately 30 days, incorporating information from 130 components of VA's electronic health records system that were previously available on disparate legacy websites into one website. Further, the officials reported that the new website increased the transparency of health care information for the veteran community, empowered veterans, and promoted competition for health care services.[53]

---

[51] The Community Development Financial Institutions Fund provides awards, for example, to institutions such as community development banks, credit unions, and other loan and venture capital funds, in order to finance activities including mortgage lending for first-time homebuyers, flexible underwriting for community facilities, and commercial loans for businesses in low-income areas. In fiscal year 2018, the fund awarded $495 million in financial assistance, loans, and bond guarantees.

[52] Access to Care is a website enabling veterans and their families to view wait times at VA facilities, compare care ratings, and review national data.

[53] In September 2017, we reported that VA provided the status of health care quality measures on its Access to Care website in a way that is accessible and easy to understand. However, we found that the agency's website only included a subset (15 out of 110) of the quality measures on VA medical centers. We therefore recommended that the agency report a broader range of health care quality measures on its website. VA concurred with our

In addition, a Defense official from the Army's Total Ammunition Management Information System[54] reported that the office had acquired infrastructure as a service in order to improve the processing and reporting of ammunition requests that the Army receives from users worldwide. Defense staff reported that, previously, they had received complaints from customers regarding system pauses and delays when entering requests for ammunition and generating reports due to legacy infrastructure. Defense officials stated that using infrastructure as a service improved system processing and reporting times—from minutes to seconds—by providing scalable technology resources that can meet worldwide performance demands. As a result, customers can more quickly enter their orders into the system.

Acquiring Cloud Services Strengthened Mission Assurance

Officials in the Office of the CIO at nine agencies reported that acquiring cloud services had allowed them to achieve greater levels of mission assurance by streamlining security resources and improving backup capabilities that were not available previously. For example, officials at Defense's North American Aerospace Defense Command and U.S. Northern Command reported that they had acquired cloud services for the Situational Awareness Geospatial Enterprise system[55] in order to improve mission assurance and address Defense cybersecurity requirements.

---

recommendation and stated that they planned to expand the quality measures reported on the website.

[54] The Total Ammunition Management Information System is a web-based system used to, among other things, validate, track, manage, and authorize ammunition requests for training and combat purposes.

[55] The Situational Awareness Geospatial Enterprise system is a global unclassified, space- based tracking system that provides situational awareness data to a variety of organizations, including local law enforcement, and federal agencies. Officials use the system during special events such as the State of the Union address and the World Series, which require additional security. In addition, the system combines air tracking data with geospatial data to track hostiles, unknowns, friendlies, and suspects with respect to the President's location at all times.

According to the officials, cloud services improved mission assurance by allowing them to more quickly correct problems such as malware and the loss of network connectivity in order to ensure the near continuous availability of data from different access points. In addition, the system required extensive storage and backup capabilities due to the need to ensure the system's data were available continuously from different access points. The officials reported that the acquisition of cloud services has reduced the costs required to maintain continuous backup and storage capabilities. They added that the system also complies with Defense requirements that investments use an approved cloud service provider. In addition, the officials said acquiring cloud services has provided the capability to scale resources, as needed, to meet demands during special events, such as the State of the Union address and the World Series, which require additional security.

Further, Federal Transit Administration officials reported that migrating two systems to the cloud had allowed the agency to enhance system security by managing access to the systems through a single portal rather than managing access to each system individually. As a result, the officials reported that they were able to shift some responsibilities of systems security management to the cloud vendor, which reduced the number of security risks and consolidated the number of security tools used. Further, according to the officials, an additional benefit is that users are required to only remember a single password rather than different passwords required for the multiple systems.

Separately, from information provided by the 15 selected agencies, we identified nine cloud computing investments that illustrate the variety of examples of benefits that had been realized by these agencies from the acquisition of cloud services. Table 7 identifies these investments and additional details regarding the nature and sources of the benefits achieved from them are profiled in appendix VI.[56]

---

[56] Additional detail on how these agencies were selected can be found in appendix I.

Table 7. Investments that benefited from cloud acquisition, as reported by selected federal agencies since 2012

Cloud investment	Agency	Managing component	Cloud operations start date
Manufacturing Support Suite	Department of Treasury	Bureau of Engraving and Printing	October 2012
National Transit Database	Department of Transportation	Federal Transit Administration	October 2014
Data to Decisions	General Services Administration	Not applicable	October 2015
Vets.gov Web Portal	Department of Veterans Affairs	Not applicable	March 2016
U.S. Trustee Program Cloud Initiative	Department of Justice	U.S. Trustee Program	March 2017
Homeland Security Information Network	Department of Homeland Security	Not applicable	July 2017
Pinyon	Department of Agriculture	U.S. Forest Service	August 2017
Weather Cloud Content Delivery Network	Department of Commerce	National Oceanic and Atmospheric Administration	September 2017
U.S. Transportation Command Cloud Program	Department of Defense	U.S. Transportation Command	January 2018

Source: GAO analysis of agency data. | GAO-19-58.

Selected Agencies Identified Key Practices That Enabled Cloud Services

In addition to the examples of significant benefits reported from acquiring cloud services, officials at the 15 agencies reported that six key practices had enabled them to realize these benefits through the successful implementation of cloud services. For example, 12 agencies reported that they implemented new governance planning activities, policies, or processes in order to help ensure that cloud acquisition efforts were managed enterprisewide. In addition, 12 agencies reported that they had reorganized the management of agency IT resources to help increase operational efficiency. Further, six agencies reported that having executive leadership involved in driving the acquisition or sponsoring efforts to use cloud services was critical for the successful adoption of cloud services across the agency. Table 8 lists these key practices and the number of agencies that reported each key practice, ranked by the number of agencies reporting the practice.

Table 8. Key practices reported by 15 selected federal agencies for helping to ensure the successful implementation of cloud computing services

Key practice activity	Number of agencies
Implemented new governance planning activities, policies, and processes	12
Modified procurement and contract oversight practices	12
Addressed changes in IT workforce needs	12
Reorganized the management of IT resources	12
Streamlined and improved IT security management	8
Engaged executive leadership to help drive acquisition efforts	6

Source: GAO analysis of agency data. | GAO-19-58.

Note: The 15 agencies were the Departments of Agriculture, Commerce, Defense, Energy, Education, Homeland Security, Justice, Labor, State, Transportation, the Treasury, and Veterans Affairs; and the General Services Administration, Small Business Administration, and Social Security Administration.

In addition, many of these six key practice areas identified by agencies are consistent with requirements outlined in FITARA[57] and recommendations from our prior work[58] made to agencies to address longstanding issues with the management of IT acquisitions and

[57] Pub. L. No. 113-291, division A, title VIII, subtitle D, § 831, 128 Stat. 3292, 3438 (Dec. 19, 2014).

[58] GAO, *Federal Chief Information Officers: Critical Actions Needed to Address Shortcomings and Challenges in Implementing Responsibilities*, GAO-18-93 (Washington, D.C.: August 2, 2018); *Information Technology Reform: Agencies Need to Improve Certification of Incremental Development*, GAO-18-148 (Washington, D.C.: November 7, 2017); *IT Workforce: Key Practices Help Ensure Strong Integrated Program Teams; Selected Departments Need to Assess Gaps*, GAO-17-8 (Washington, D.C.: November 30, 2016); *Information Technology: Additional OMB and Agency Actions Needed to Ensure Portfolio Savings Are Realized and Effectively Tracked,* GAO-15-296 (Washington, D.C.: April 16, 2015) and *IT Dashboard: Agencies Need to Fully Consider Risks When Rating Their Investments,* GAO-16-494 (Washington, D.C.: June 2, 2016).

operations. Specifically, we previously have noted the importance of strengthening the authority of CIOs, improving the portfolio review process and the transparency of major investment data, ensuring the use of incremental development methodologies, and updating human capital plans.

Selected Agencies Implemented New Policies and Processes to Guide Governance of Cloud Acquisition

Officials in the Office of the CIO at 12 agencies reported that they had implemented new governance activities or drafted new policies and processes to help ensure the successful implementation of cloud services. For example, SSA officials reported that they had drafted several new policies to simplify the management of cloud resources and provide better oversight for new cloud service acquisitions. Specifically, the officials reported that the new policies established a request-and- approval governance process to address which staff can initiate cloud solutions and what types of projects can receive funding.

In addition, Energy officials reported that they had formalized policies and governance processes on how to perform cloud migrations, including establishing a documented, repeatable process to help offices migrate to cloud services more efficiently. Further, Treasury officials reported that they had focused on strengthening cloud-governance activities, including planning and identifying requirements, because changes to enterprise cloud systems may impact multiple programs. As a result, these officials reported that they had implemented a cross-cutting steering committee to help better plan and assess the impact of changes to enterprise cloud systems that support multiple programs.

Selected Agencies Modified Procurement and Contract Oversight Practices to Strengthen Cloud Acquisition Processes

Officials in the Office of the CIO at 12 agencies reported that they had modified their procurement and contract oversight practices in order to accommodate the differences in how cloud services are acquired from traditional acquisitions. For example, Commerce officials emphasized the importance of developing standardized requirements to ensure that when bureaus award contracts, they use standardized language. The officials stated that these requirements help to ensure that contracts with cloud service providers are comprehensive, legally adequate, and include specific details regarding all of the activities the agency will need the contractor to perform.

In addition, officials of the U.S. Trustee Program at Justice reported that they had used existing project and financial management resources to monitor the use of cloud services and associated spending to help control costs and ensure the accuracy of cloud vendor charges. For example, the officials reported that the program used the cloud vendor's administrative and business intelligence tools to create reports to verify cloud charges. Also, Labor officials reported that they had worked with the agency's acquisition team to ensure the agency is only billed for its actual cloud usage. This required the agency to transition from a fixed-price contract model to a time and materials-based contract model, which included a clause that limited the maximum costs the agency would have to pay for cloud services.

Selected Agencies Addressed Changes in Workforce Needs for Managing Cloud Services

Officials in the Office of the CIO at 12 agencies reported that they had taken several steps to address changes in workforce needs for managing cloud services. Specifically, these officials reported that they had conducted inventories of staff skills, transitioned staff into new roles, and ensured that staff acquired training. For example, VA officials reported that they had conducted a staff skills inventory to identify future IT workforce training needs and transition staff from managing legacy technologies to managing cloud services. In addition, Energy officials reported that they were preparing staff to transition from managing data center resources to managing the agency's service level agreements with cloud providers. The officials reported that moving to cloud services allowed staff to spend more time improving existing applications and identifying other efforts to innovate IT services rather than managing on-premise infrastructure. Further, a Defense official lead for cloud computing in the Navy's Office of the CIO reported that the Navy had developed an enterprise cloud working group consisting of key members from major offices and security groups to help determine the appropriate training and certification needs for staff and to conduct training seminars. In addition, SBA officials said that the agency took advantage of a contract option offered by the cloud vendor to acquire free cloud classes and training, thereby avoiding the need to spend approximately $380,000 on training.

Selected Agencies Reorganized the Management of IT Resources to Increase Operational Efficiency

Office of the CIO officials at 12 agencies reported that acquiring cloud services had led them to change how they organized and managed the agency's IT resources. For example, GSA officials reported that they had transitioned from letting individual components within the agency acquire their own application to using an enterprise approach in which software as a service applications are acquired and made available to the entire agency. As a result, the officials reported that this approach allows the agency to further optimize their software purchases and improve their monitoring and tracking of software application usage enterprise-wide.

In addition, officials at the Agricultural Research Service reported that acquiring software as a service had promoted opportunities to share customizations of the acquired software between Agriculture's components rather than having each component develop a separate customization. Specifically, these officials reported that they were able to take a software feature developed by another Agriculture component and implement it for their customer service portal, rather than having to develop it themselves.

Lastly, Education officials reported that they were in the process of beginning an assessment to consolidate the agency's existing cloud services across federal and commercial environments. The officials said that they hoped to reduce the number of commercial cloud providers used from twenty-five to eight, and to consolidate two of the agency's cloud environments into a single environment within the next 3 years.

Selected Agencies Streamlined Cloud Services to Address Security Needs in a More Efficient Manner

Office of the CIO officials at eight agencies reported that they were able to streamline the management of IT security by leveraging cloud services. For example, SBA officials reported that they used security tools from their cloud vendor in order to meet DHS's requirements for continuous diagnostics and mitigation and improve the agency's security posture. The officials reported that they had performed a requirements analysis and found that, compared to acquiring costly hardware solutions to manage this capability internally, their existing cloud vendor provided security capabilities that actually exceeded DHS's recommended continuous diagnostics and mitigation requirements. As a result, SBA adopted the cloud vendor's security tools and avoided $300,000 in initial hardware purchases, as well as subsequent hardware technology refreshes every 3 years.

In addition, GSA officials reported that choosing a FedRAMP-approved cloud service provider had expedited the agency's adoption of cloud services. Specifically, the agency did not have to visit and review each vendor's facility as part of the vendor approval process, which shortened the time frame needed to approve a system for use. The officials also reported that using cloud services streamlined the deployment process for new systems because using a cloud platform that had previously been granted the authority to operate[59] allowed the agency to avoid undertaking a separate authorization process, which saved time and resources.

[59] National Institute of Standards and Technology, *Guide for Applying the Risk Management Framework to Federal Information Systems: A Security Life Cycle Approach*, Special Publication 800-37 (Revision 1, updated June 5, 2014), defines the authority to operate as the official management decision given by a senior organizational official to authorize operation of an information system and to explicitly accept the risk to organizational operations (including mission, functions, image, or reputation), organizational assets, individuals, other organizations, and the nation based on the implementation of an agreed-upon set of security controls.

Selected Agencies Engaged Executive Leadership Support during Cloud Acquisition to Help Ensure Successful Implementation

Officials in the Office of the CIO at six agencies reported that having executive leadership involved in driving acquisitions or sponsoring efforts to use cloud services was critical to the successful adoption of cloud services across the agency. For example, SBA officials reported that their agency CIO's commitment to acquiring cloud services and the deputy CIO's attendance at daily cloud meetings were critical for the successful adoption of cloud services at the agency. Similarly, Energy officials reported that the agency had established a team with representatives from offices of the CIO, chief financial officer, and chief acquisition officer, to coordinate IT expenditures, including cloud investments, across the agency.

Further, Defense officials from the U.S. Transportation Command reported that establishing a cloud center of excellence team that reported directly to the Commander of U.S. Transportation Command had empowered the team to engage directly with users to help break down barriers that might impact the migration to cloud services. In addition, the officials said that the team helped streamline the processes—specifically, the design, contracting, funding, transition planning, and implementation processes—necessary for the successful migration of all of the command's systems to the cloud.

CONCLUSION

Since 2011, when OMB began requiring agencies to adopt a Cloud First strategy, agencies have made progress in implementing cloud services and, in doing so, have saved hundreds of millions of dollars and realized notable benefits. However, six agencies still lack guidance for assessing IT investments for cloud services and 12 agencies still have not performed assessments for a number of their IT investments. In addition, all of the agencies in our review do not have sufficient mechanisms or approaches to track and report the savings data associated with these cloud initiatives.

Although agencies have reported spending $1 billion or more on cloud services in just the past 2 years, and identified hundreds of millions of dollars in related savings, these figures are not consistently reported. To its credit, beginning in fiscal year 2019, OMB will require more accuracy and granularity in how agencies report cloud investment spending data. However, there has not been a corresponding effort to improve the reporting of cloud savings data. An important aspect to the success of key OMB cloud initiatives, like Cloud Smart and the associated drive for greater agency adoption of cloud services, will be the ability for key stakeholders to access complete information on the savings that agencies are achieving under these efforts.

Recommendations for Executive Action

We are making a total of 35 recommendations—1 recommendation to OMB and 34 recommendations to the 16 agencies in our review.

The Director of the Office of Management and Budget should require agencies to explicitly report, at least on a quarterly basis, the savings and cost avoidance associated with cloud computing investments. (Recommendation 1)

The Secretary of Agriculture should ensure that the CIO of Agriculture completes an assessment of all IT investments for suitability for migration to a cloud computing service, in accordance with OMB guidance. (Recommendation 2)

The Secretary of Agriculture should ensure that the CIO of Agriculture establishes a consistent and repeatable mechanism to track savings and cost avoidances from the migration and deployment of cloud services. (Recommendation 3)

The Secretary of Commerce should ensure that the CIO of Commerce establishes a consistent and repeatable mechanism to track savings and cost avoidances from the migration and deployment of cloud services. (Recommendation 4)

The Secretary of Defense should ensure that the CIO of Defense completes an assessment of all IT investments for suitability for migration to a cloud computing service, in accordance with OMB guidance. (Recommendation 5)

The Secretary of Defense should ensure that the CIO of Defense establishes a consistent and repeatable mechanism to track savings and cost avoidances from the migration and deployment of cloud services. (Recommendation 6)

The Secretary of Education should ensure that the CIO of Education establishes guidance on assessing new and existing IT investments for suitability for cloud computing services, in accordance with OMB guidance. (Recommendation 7)

The Secretary of Education should ensure that the CIO of Education completes an assessment of all IT investments for suitability for migration to a cloud computing service, in accordance with OMB guidance. (Recommendation 8)

The Secretary of Education should ensure that the CIO of Education establishes a consistent and repeatable mechanism to track savings and cost avoidances from the migration and deployment of cloud services. (Recommendation 9)

The Secretary of Energy should ensure that the CIO of Energy updates the agency's guidance on assessing IT investments for suitability for cloud computing services to include a requirement to assess new acquisitions for these services. (Recommendation 10)

The Secretary of Energy should ensure that the CIO of Energy completes an assessment of all IT investments for suitability for migration to a cloud computing service, in accordance with OMB guidance. (Recommendation 11)

The Secretary of Energy should ensure that the CIO of Energy establishes a consistent and repeatable mechanism to track savings and cost avoidances from the migration and deployment of cloud services. (Recommendation 12)

The Secretary of Health and Human Services should ensure that the CIO of HHS establishes guidance on assessing new and existing IT investments for suitability for cloud computing services, in accordance with OMB guidance. (Recommendation 13)

The Secretary of Health and Human Services should ensure that the CIO of HHS completes an assessment of all IT investments for suitability for migration to a cloud computing service, in accordance with OMB guidance. (Recommendation 14)

The Secretary of Health and Human Services should ensure that the CIO of HHS establishes a consistent and repeatable mechanism to track savings and cost avoidances from the migration and deployment of cloud services. (Recommendation 15)

The Secretary of Homeland Security should ensure that the CIO of DHS completes an assessment of all IT investments for suitability for migration to a cloud computing service, in accordance with OMB guidance. (Recommendation 16)

The Secretary of Homeland Security should ensure that the CIO of DHS establishes a consistent and repeatable mechanism to track savings and cost avoidances from the migration and deployment of cloud services. (Recommendation 17)

The Attorney General of the United States should ensure that the CIO of Justice completes an assessment of all IT investments for suitability for migration to a cloud computing service, in accordance with OMB guidance. (Recommendation 18)

The Attorney General of the United States should ensure that the CIO of Justice establishes a consistent and repeatable mechanism to track savings and cost avoidances from the migration and deployment of cloud services. (Recommendation 19)

The Secretary of Labor should ensure that the CIO of Labor updates the agency's guidance on assessing IT investments for suitability for cloud computing services to include a requirement to assess existing investments for these services. (Recommendation 20)

The Secretary of Labor should ensure that the CIO of Labor completes an assessment of all IT investments for suitability for migration to a cloud computing service, in accordance with OMB guidance. (Recommendation 21)

The Secretary of Labor should ensure that the CIO of Labor establishes a consistent and repeatable mechanism to track savings and cost avoidances from the migration and deployment of cloud services. (Recommendation 22)

The Secretary of State should ensure that the CIO of State establishes a consistent and repeatable mechanism to track savings and cost avoidances from the migration and deployment of cloud services. (Recommendation 23)

The Secretary of the Treasury should ensure that the CIO of Treasury completes an assessment of all IT investments for suitability for migration to a cloud computing service, in accordance with OMB guidance. (Recommendation 24)

The Secretary of the Treasury should ensure that the CIO of Treasury establishes a consistent and repeatable mechanism to track savings and cost avoidances from the migration and deployment of cloud services. (Recommendation 25)

The Secretary of Transportation should ensure that the CIO of Transportation establishes guidance on assessing new and existing IT investments for suitability for cloud computing services. (Recommendation 26)

The Secretary of Transportation should ensure that the CIO of Transportation completes an assessment of all IT investments for suitability for migration to a cloud computing service, in accordance with OMB guidance. (Recommendation 27)

The Secretary of Transportation should ensure that the CIO of Transportation establishes a consistent and repeatable mechanism to track savings and cost avoidances from the migration and deployment of cloud services. (Recommendation 28)

The Secretary of Veterans Affairs should ensure that the CIO of VA completes an assessment of all IT investments for suitability for migration to a cloud computing service, in accordance with OMB guidance. (Recommendation 29)

The Secretary of Veterans Affairs should ensure that the CIO of VA establishes a consistent and repeatable mechanism to track savings and cost avoidances from the migration and deployment of cloud services. (Recommendation 30)

The Administrator of General Services should ensure that the CIO of GSA establishes a consistent and repeatable mechanism to track savings and cost avoidances from the migration and deployment of cloud services. (Recommendation 31)

The Administrator of the Small Business Administration should ensure that the CIO of SBA establishes a consistent and repeatable mechanism to track savings and cost avoidances from the migration and deployment of cloud services. (Recommendation 32)

The Commissioner of the Social Security Administration should ensure that the CIO of SSA updates the agency's guidance on assessing IT investments for suitability for cloud computing services to include a requirement to assess existing investments for these services. (Recommendation 33)

The Commissioner of the Social Security Administration should ensure that the CIO of SSA completes an assessment of all IT investments for suitability for migration to a cloud computing service, in accordance with OMB guidance. (Recommendation 34)

The Commissioner of the Social Security Administration should ensure that the CIO of SSA establishes a consistent and repeatable mechanism to track savings and cost avoidances from the migration and deployment of cloud services. (Recommendation 35)

AGENCY COMMENTS AND OUR EVALUATION

We provided a draft of this chapter to OMB and the 16 agencies for their review and comment. In response, 14 agencies provided comments stating that they agreed with our recommendations; one agency stated that it agreed with one recommendation but disagreed with another; and two agencies did not state whether they agreed or disagreed with the recommendations. In addition, multiple agencies provided technical comments, which we incorporated into the report, as appropriate.

The following 14 agencies agreed with our recommendations:

- In written comments from Commerce, Education, Energy, HHS, DHS, State, Transportation, VA, and GSA, these agencies stated that they agreed with the recommendations directed to them. In addition, each of the agencies indicated that it planned, or already had begun taking actions, to address the recommendations. The agencies' comments are reprinted in appendixes VII through XV, respectively.
- In emails received from Agriculture's Director of Strategic Planning, Policy, Egovernment and Audits in the Office of the CIO on February 11, 2019, and from Justice's Audit Liaison Specialist in the Internal Review and Evaluation Office on February 15, 2019, both of these departments stated that they agreed with our recommendations.
- In written comments from Labor, the department stated that it agreed with our recommendations. The department also described actions taken to address our recommendation that it update its guidance on assessing IT investments for suitability for cloud computing services to include a requirement to assess existing investments for these services. Specifically, Labor stated that it had taken steps to ensure that its agencies included an assessment of cloud computing suitability as they moved forward with their investments and that this process had been integrated into Labor's budget process. We followed up with the department and obtained a copy of Labor's guidance.

  However, in examining this guidance,[60] we found it to be the same as what Labor had previously provided to us during the course of our audit. Further, as we mentioned earlier regarding our analysis of the department's guidance for assessing investments for suitability for cloud services, Labor had required existing investments that were already using cloud services to migrate to the department's new consolidated cloud environment; however, it did not require

[60] Department of Labor, *DOL OCIO Cloud Consolidation Initiative (DC^{2}I) Application and Service Migration to DOL Cloud, version 1.5* (March 2018).

existing systems not using cloud services to be assessed for these services. Without receiving any additional information from the department that supported its actions to address our recommendation prior to this chapter's issuance, we believe our recommendation to Labor is still appropriate. The department's comments are reprinted in appendix XVI.

- In written comments from SBA, the agency agreed with our recommendation. Also, in additional comments sent by a GAO liaison in the Office of Congressional and Legislative Affairs via email on March 11, 2019, SBA provided updated information regarding the benefits that the agency had realized from using cloud services for its system that was profiled in appendix VI of the draft report. Specifically, SBA officials in charge of the system provided a revised list of realized benefits from the cloud services. However, the officials did not provide any supporting documentation regarding the revised benefits; therefore, we were not able to validate the revised list of benefits prior to the issuance of this chapter. As a result, we removed the profile from the report in order to be consistent with our methodology for reporting examples of systems that had realized benefits from the acquisition of cloud services, and notified SBA of this decision. SBA's comments are reprinted in appendix XVII.
- In written comments from SSA, the agency agreed with our recommendations. Also, in additional comments that a management analyst from the office of the Deputy Commissioner for Analytics, Review and Oversight provided via email on January 11, 2019, SSA noted that the agency was in the process of acquiring a new vendor solution for its system that was profiled in appendix VI of the draft report. However, during further discussion with SSA officials in charge of the system on January 17, 2019, the officials confirmed that the agency had not yet identified all of the potential benefits related to the use of cloud services as a result of a change in their vendor solution. Thus, we removed the profile from our report in order to be consistent with our methodology for reporting examples

of systems that had realized benefits from the acquisition of cloud services, and notified SSA of this decision. SSA's comments are reprinted in appendix XVIII.

One agency agreed with one recommendation and disagreed with a second recommendation:

- Defense provided written comments in which it agreed with our recommendation to complete an assessment of all IT investments for suitability for migration to a cloud computing service. However, the agency did not agree with our recommendation that it establish a mechanism to track savings and cost avoidances from the migration and deployment of cloud services. Specifically, Defense stated that it did not agree with our recommendation because there was no standard, consistent way to capture such savings or cost avoidance. The department stated that it would work with OMB on whether or how to collect such information, and, if practical, report such information in accordance with OMB guidance.
  However, as we noted in our report, for the past 6 years, OMB has required agencies to report on savings and cost avoidances from implementing IT reform initiatives, including savings realized from the migration to cloud services. Tracking savings and cost avoidances for cloud initiatives is important in order to ensure that Defense is effectively managing and overseeing its cloud initiatives. In addition, it is essential that OMB and Congress have sufficient data to see the results of Defense's cloud initiatives and understand whether the department is achieving savings using cloud services. Consequently, we believe our recommendation to track savings and cost avoidances from the migration and deployment of cloud services is still warranted. The department's comments are reprinted in appendix XIX.

Finally, we received comments via email from Treasury's Supervisory IT Specialist in the Office of the CIO on February 22, 2019, and OMB's

Liaison to GAO on February 25, 2019. In these comments these two agencies did not state whether they agreed or disagreed with the recommendations that we directed to them.

As agreed with your offices, unless you publicly announce the contents of this chapter earlier, we plan no further distribution until 30 days from the report date. At that time, we will send copies of this chapter to the Director of the Office of Management and Budget, the Secretaries and agency heads of the departments and agencies in this chapter, and other interested parties.

Carol C. Harris
Director, Information Technology
Acquisition Management Issues

List of Requesters

The Honorable Elijah E. Cummings
Chairman

The Honorable Jim Jordan
Ranking Member
Committee on Oversight and Reform
House of Representatives

The Honorable Gerry Connolly
Chairman

The Honorable Mark Meadows
Ranking Member
Subcommittee on Government Operations
Committee on Oversight and Reform
House of Representatives

The Honorable Will Hurd
House of Representatives

The Honorable Robin L. Kelly
House of Representatives

APPENDIX I: OBJECTIVES, SCOPE, AND METHODOLOGY

Our objectives for this engagement were to (1) evaluate selected agencies' progress in implementing cloud services, (2) review the extent to which selected agencies have increased spending on cloud services and achieved cost savings or avoidances, and (3) describe examples of cloud investments with significant or notable benefits that have been identified by selected agencies.

For this review, we selected a sample of agencies based on the size of their total information technology (IT) budget for fiscal year 2017. Specifically, we categorized each of the 24 *Chief Financial Officers Act* agencies[61] by the size of its IT budget: large (more than $3 billion), medium ($1 billion to $3 billion), and small (less than $1 billion), as reported on the Office of Management and Budget's (OMB) IT Dashboard[62] for 2017. We then selected up to six agencies with the largest budgets from each budget category.

Using these criteria, we selected 16 agencies, including seven agencies from our prior work on cloud services conducted in 2012 and 2014.[63]

---

[61] The 24 major agencies listed in the *Chief Financial Officers Act of 1990* are the Departments of Agriculture, Commerce, Defense, Education, Energy, Health and Human Services, Homeland Security, Housing and Urban Development, the Interior, Justice, Labor, State, Transportation, the Treasury, and Veterans Affairs; the Environmental Protection Agency, General Services Administration, National Aeronautics and Space Administration, National Science Foundation, U.S. Nuclear Regulatory Commission, Office of Personnel Management, Small Business Administration, Social Security Administration, and U.S. Agency for International Development. 31 U.S.C. § 901(b).

[62] OMB, IT Dashboard, 2017 (https://itdashboard.gov).

[63] GAO, *Cloud Computing: Additional Opportunities and Savings Need to Be Pursued*, GAO-14-753 (Washington, D.C.: Sept. 25, 2014); and *Information Technology Reform: Progress*

These agencies were the Department of Agriculture (Agriculture), Department of Commerce (Commerce), Department of Defense (Defense), Department of Education (Education), Department of Energy (Energy), Department of Health and Human Services (HHS), Department of Homeland Security (DHS), Department of Justice (Justice), Department of Labor (Labor), Department of State (State), Department of Transportation (Transportation), Department of the Treasury (Treasury), Department of Veterans Affairs (VA), General Services Administration (GSA), Small Business Administration (SBA), and Social Security Administration (SSA).

To address the first objective, we obtained and analyzed IT Dashboard data related to the 16 selected agencies' use of cloud services for fiscal years 2016 through 2018, and their projected use in 2019. We chose to begin with fiscal year 2016 because we had previously reported on federal agencies' use of cloud services through fiscal year 2014 and fiscal year 2015 data was not available. Specifically, the Dashboard includes agency responses to a cloud-related question from OMB's capital planning guidance.[64] The question asks whether a cloud alternative was evaluated for the investment, or components or systems within the investment. We reviewed agency responses that were submitted for fiscal years 2016 through 2019 as part of the annual budget submission process in order to determine whether a specific investment was using cloud services.

During this 4-year period, OMB made changes to the options that agencies were required to choose from which indicated whether an investment was using cloud services. For OMB's capital planning guidance for fiscal years 2016 and 2017, we selected responses that indicated that the agency "had evaluated a cloud alternative and chose a cloud alternative" with a particular cloud deployment model.[65] In addition, for

Made but Future Cloud Computing Efforts Should be Better Planned, GAO-12- 756 (Washington, D.C.: July 11, 2012).

[64] OMB, *FY 2016 IT Budget–Capital Planning Guidance* (Washington, D.C.: June 27, 2014); *FY 2017 IT Budget–Capital Planning Guidance* (Washington, D.C.: June 16, 2015); *FY 2018 IT Budget–Capital Planning Guidance* (Washington, D.C.: June 30, 2016); *and FY 2019 IT Budget–Capital Planning Guidance* (Washington, D.C.: Aug. 1, 2017).

[65] OMB required agencies to select one of the following options: (1) the agency evaluated a cloud alternative and chose a cloud alternative with a public deployment model for some or all of

OMB's capital planning guidance for fiscal years 2018 and 2019, we selected responses that indicated the "investment or a portion of the investment is leveraging cloud computing."[66] We then determined the total number of investments using cloud services and calculated the percentage of investments using these services based on the total number of reported investments by each agency for each fiscal year.

To ensure the accuracy and completeness of the selected agencies' data on the use of cloud services, we downloaded this data from the IT Dashboard on October 3, 2017, March 7, 2018, and October 9, 2018. We took this step because agencies may update their data on a quarterly basis throughout the fiscal year. In addition, we presented the results of our analysis to officials in charge of cloud services within the Office of the Chief Information Officer (CIO) at each selected agency. We asked these officials to verify the completeness and accuracy of this data and provide any updates as appropriate. Officials at all 16 agencies confirmed the total number of investments using cloud services for fiscal years 2016 through 2018 and their projected use for fiscal year 2019. Based on these steps, we determined that these data were sufficiently reliable to report on agencies' progress in using cloud services.

In addition, we compared each selected agency's cloud guidance to OMB's Cloud First guidance. We interviewed Office of the CIO officials in charge of cloud services at each agency regarding their guidance. In addition, we interviewed OMB staff from the Office of E-Government and Information Technology regarding its guidance. Because of the wide

---

the investment; (2) the agency evaluated a cloud alternative and chose a cloud alternative with a private deployment model for some or all of the investment; (3) the agency evaluated a cloud alternative and chose a cloud alternative with a community deployment model for some or all of the investment; (4) the agency evaluated a cloud alternative and chose a cloud alternative with hybrid/mixed deployment models for some or all of the investment. This option also applies if the investment contains more than one cloud alternative deployment model; (5) the agency evaluated a cloud alternative but did not choose a cloud alternative for any of the investment; and (6) the agency did not evaluate a cloud alternative but plans to evaluate a cloud alternative by the end of the budget year.

[66] OMB required agencies to select from one of the following options: (1) this investment or a portion of this investment is leveraging cloud computing; (2) this investment is migrating to the cloud; (3) this investment is considering cloud computing; (4) cloud computing has not been considered; (5) cloud computing is not applicable for any portion of this Investment; and (6) cloud computing has been considered but was not selected.

variety of responses and documents we received from the agencies related to their guidance for assessing investments for cloud computing services, we conducted a content analysis of the information in order to determine compliance with OMB's guidance. In doing so, team members individually reviewed agencies' responses and documents and assigned them to various categories and subcategories. Team members then compared their categorization schemes, discussed the differences, and reached agreement on the final characterization of compliance with OMB guidance. In cases where agencies provided multiple policies or documents, we followed up to clarify which portions were considered by the agency to support the requirement to assess all investments for cloud services.

In analyzing whether the agencies' guidance on assessing investments for cloud services met OMB criteria, we assessed whether the guidance clearly identified a requirement for evaluating both new and existing investments for cloud services. Agencies found to not have guidance which clearly defined the assessment process were evaluated as such for one of two reasons: either the agency's formal guidance did not completely address our assessment criteria or the agency's guidance had not yet been established or finalized.

In analyzing whether agencies had met OMB's requirement to evaluate each investment for cloud services, we assessed the number of investments that had completed assessments based on the fiscal year 2019 budget submission. Agencies found not to have met the requirement were evaluated as such if the agency had 10 or more investments that had not yet been evaluated for cloud services. We set this threshold based on a reasonable interpretation of the intent of OMB's guidance requiring assessments of all investments.

For our second objective, we obtained and analyzed IT Dashboard data related to the 16 agencies' spending on cloud services for fiscal years 2015 through 2018. We chose to begin with fiscal year 2015 because we had previously reported on federal agencies' spending on cloud services through fiscal year 2014. Agencies report actual spending costs by fiscal year on the IT Dashboard as part of the next fiscal year reporting. To determine actual cloud spending costs for each fiscal year, we used agency spending data reported each subsequent fiscal year (from fiscal years 2017 through 2018) as of October 5, 2018.

In addition, we administered a data collection instrument to each of the 16 agencies to obtain and analyze spending and savings data by the 16 selected agencies for fiscal years 2014 through 2018 and plans for future planned costs. We requested that these agencies provide spending and savings data broken down by investment, as OMB only requires federal agencies to report total spending by cloud deployment model on the IT Dashboard, and agencies were not required to identify whether any reported savings were cloud-related. This instrument was administered from November 2017 to January 2018.

In the data collection instrument, we asked the selected agencies to complete information on each of their cloud investments, including the title of the application or system leveraging cloud, the cloud deployment and service models, and the associated cloud spending and net cloud savings or avoidances from fiscal year 2014 through fiscal year 2018 and beyond, as agencies generally submitted data on planned spending for one or more fiscal years beyond 2018. Due to the varied scale of cloud implementation efforts ongoing at these agencies, we asked agencies to only provide all applications, systems, or investments leveraging cloud with total life-cycle costs of $1 million or more. We also asked agencies to provide spending and savings or cost avoidances figures in whole numbers in order to avoid errors in rounding numbers when we calculated the reported figures in millions.

We took the following steps to help ensure the reliability of the data we collected. First, to minimize errors that might occur from respondents interpreting our instrument differently from our intended purpose, we reviewed the data collection instrument with agency officials who would be completing the instrument during meetings in October and November 2017. Second, we reviewed the completed spreadsheets to identify missing data or other errors, and consulted with our data quality expert about these issues as appropriate.

All agencies completed the data collection instrument by May 2018. For those agencies that provided rounded (rather than exact) spending or savings figures, we recalculated the data into whole numbers and confirmed our calculations with the agencies. In addition, one agency broke down its savings data into savings and cost avoidances; we combined these reported figures for each investment and, after consultation with a GAO data subject matter expert, confirmed with all the other agencies in our review that their information on savings also included cost avoidances.

We also reviewed the associated notes regarding agencies' qualifications of the provided data and followed up with agency officials to clarify the responses as appropriate. These notes included information on whether certain spending or savings data were unavailable, whether certain costs were excluded from the spending information provided to us or whether there were other qualifications of the provided data.

Lastly, we presented the results of our analysis of IT Dashboard data and the data obtained from the data collection instrument to each of the selected agencies between June and August 2018. We asked the agencies to verify the completeness and accuracy of these data and provide any updates as appropriate. All 16 agencies provided updated information regarding the list of investments using cloud services with life- cycle costs of $1 million or more and six agencies (Agriculture, Commerce, Justice, Transportation, Treasury, and VA) provided updated information related to spending and savings for these investments, which we have incorporated as appropriate. Based on the measures we took to ensure the reliability of the data provided by the agencies and reported on the IT Dashboard, we determined that the data were sufficiently reliable for the purpose of this chapter.

For the third objective, we obtained and reviewed available documentation discussing examples of cloud computing investments reported by the selected agencies as having produced notable benefits and key practices that ensure the effort was successful. We also interviewed officials from the Office of the CIO and other components in charge of cloud services regarding these benefits.

In order to develop our list of questions for these meetings, we first conducted research to identify the range of benefits that could be achieved from acquiring cloud services. We reviewed OMB, GSA, and CIO Council cloud guidance; our prior work;[67] and key leading cloud practices from GSA's Federal Cloud Computing Center of Excellence. Based on this work, we developed a list of seven key areas of benefits: (1) improving efficiency and operations; (2) promoting agility and responsiveness; (3) achieving business growth; (4) reducing cost; (5) meeting regulatory requirements; (6) enhancing customer experience; and (7) ensuring mission outcomes. During meetings with agency officials in the Office of the CIO and other components in charge of cloud services, we asked officials whether they had identified any significant or notable benefits in these seven areas. As these seven areas might not represent all potential benefits, we also asked officials to describe any additional benefits not included in these areas.

In addition, as part of these meetings, we asked officials from the Office of the CIO at each selected agency to identify up to three examples of investments that benefited from the acquisition of cloud services. We asked agencies to exclude examples of email deployments to the cloud to ensure a wider variety of examples of investments with benefits. Fifteen of the 16 agencies in our review identified at least one or more examples of cloud investments that had produced significant or notable benefits, while one agency—HHS—reported that it did not have any such examples because it did not have any completed migration efforts. Because of the open-ended nature of the 15 agencies' responses to our questions, we conducted a content analysis of the information we received in order to identify and summarize the benefits and key practices that were identified by the 15 agencies. We reviewed the benefits and key practices reported by the agencies and grouped them using the seven key benefit areas that our prior research had identified. We discussed the groupings of the reported benefits, and reached agreement on these categories. We grouped the benefit categories together based on commonalities such as purpose, impact, and capabilities, and summarized the benefits reported. Based on

---

[67] GAO-12-756 and GAO-14-753.

discussion, we confirmed a list of benefits and key practices and totaled the number of agencies that reported each of these.

In addition, to select systems or investments to profile, we reviewed the 34 examples provided by the 15 agencies and narrowed the list to 11 examples. We selected these examples using the following factors: the type of system, whether the system supported the mission or business operations of the agency or component, and the availability of information related to the benefits achieved from acquiring cloud services. In doing so, we sought to have a mix of systems that provided mission critical services to the agency or the public, illustrated a range of cloud computing benefits, and included detailed information on the benefits achieved from using cloud services.

In technical comments received on a draft of this chapter, two agencies provided new information regarding the use of cloud services for their systems that were profiled in appendix VI of the draft report. Based on the additional information provided by the two agencies, we determined there was no longer sufficient detail regarding what benefits were realized for these systems. Therefore, we removed the two agencies' profiled examples from the report in order to be consistent with our methodology for reporting examples of systems that had realized benefits from the acquisition of cloud services. We then notified both agencies of this decision.

We conducted this performance audit from September 2017 to April 2019 in accordance with generally accepted government auditing standards. Those standards require that we plan and perform the audit to obtain sufficient, appropriate evidence to provide a reasonable basis for our findings and conclusions based on our audit objectives. We believe that the evidence obtained provides a reasonable basis for our findings and conclusions based on our audit objectives.

APPENDIX II: STATUS OF CLOUD-RELATED RECOMMENDATIONS IN THE 2017 AMERICAN TECHNOLOGY COUNCIL REPORT

In May 2017, the Administration established the American Technology Council to help transform and modernize federal IT and how the government uses and delivers digital services.

Subsequently, in December 2017, the American Technology Council issued a Report to the President on Federal IT Modernization[68] and made eight cloud computing- related recommendations that are relevant to the focus of our review.

Table 9. Status of recommendations related to advancing cloud computing within the federal government in the report to the president on Information Technology (IT) modernization, according to Office of Management and Budget (OMB), as of July 2018

Recommendation number	Responsible party	Required action	Current recommendation status according to OMB
13	OMB	Submit a data call to agencies requesting submission of both in-progress and pending projects for cloud migration.	OMB completed this action in December 2017.
14	Federal Agencies	Propose a cloud migration plan that highlights needed changes to requisite policies and capabilities to facilitate faster migration.	Agencies submitted plans to OMB in January 2018. OMB was in the process of verifying the information provided and planned to send out a second data call in July 2018.[a]

[68] White House American Technology Council, *Report to the President on Federal IT Modernization* (Washington, D.C.: Dec. 13, 2017).

Recommendation number	Responsible party	Required action	Current recommendation status according to OMB
25	OMB	Issue data call that will have agencies identify systems that may be ready for cloud migration and can be migrated securely but have not yet migrated due to perceived or encountered difficulties.	OMB completed this action in December 2017.
26	Federal Agencies	Respond to OMB data call regarding systems that may be ready for cloud migration and can be migrated securely but have not yet migrated due to perceived or encountered difficulties.	Agencies have provided data on their systems to OMB.
27	OMB and General Services Administratio n	Review the impediments to moving to the cloud outlined by agencies and prioritize an infusion of technical talent, capital, and the updating of security policy (developed iteratively to solve agency-specific issues) as needed to enable prioritized cloud migrations.	OMB has completed its review and will be addressing these issues in the upcoming Cloud Computing Strategy to be issued in the next few months. The new draft strategy for comment issued in September 24, 2018 identified a range of action items that will be undertaken over the next eighteen months, including the release of new security-related policies.
29	OMB, in coordination with Department of Homeland Security and other Federal partners	Update the Federal Cloud Computing Strategy, which will provide additional guidance to agencies on the most impactful use cases for cloud adoption and how best to conduct appropriate operational security in cloud environments.	OMB issued a draft strategy for comment on September 24, 2018.

Table 9. (Continued)

Recommendation number	Responsible party	Required action	Current recommendation status according to OMB
30	OMB	Conduct a thorough review of all relevant policies pertaining to IT modernization, cloud migration, infrastructure consolidation, and shared services, among others, and initiate revisions, rescissions, or other rapid policy updates that may improve the ability of agencies to modernize effectively, securely, and efficiently. If necessary, OMB is to issue further guidance that will augment and enhance existing Federal technology and information security policy.	OMB officials reported that this task was ongoing. The officials stated that certain policy issues are to be addressed in the forthcoming update to the Cloud Computing Strategy.The new draft strategy, which was issued for comment on September 24, 2018, identified a range of action items that are to be undertaken over the next 18 months, including the release of new policies and guidance related to cloud computing services.
36	OMB	Establish a comprehensive strategy for driving the accelerated migration of agency email and collaboration tools to the cloud for departments and agencies that still have not adopted cloud- based email.	Information about the progress of email migration is to be released as part of the updated Cloud Computing Strategy. The new draft strategy, which was issued for comment on September 24, 2018, identified a range of action items that are to be undertaken over the next 18 months, but there was no specific mention of agency email in the list.

Source: GAO analysis of OMB data. | GAO-19-58.

[a]OMB was unable to provide an updated status of this planned action in time to be included in the final report.

Table 9 outlines the cloud-related recommendations contained in the report and the current status of these recommendations as of July 2018, according to Office of Management and Budget (OMB) staff from the Office of E-Government and Information Technology.

APPENDIX III: SELECTED AGENCY CLOUD INVESTMENTS FOR FISCAL YEARS 2016 THROUGH 2019

Table 10. Cloud computing investments and Information Technology (IT) investments, as reported on the IT dashboard by 16 selected agencies, for fiscal years 2016-2019 (projected)

Agency	Cloud investments/ IT investments, fiscal year 2016	Cloud investments/ IT investments, fiscal year 2017	Cloud investments/ IT investments, fiscal year 2018	Cloud investments/ IT investments, fiscal year 2019
Total number				
Department of Agriculture	21/216	23/204	46/233	53/225
Department of Commerce	22/143	22/142	35/153	45/167
Department of Defense	146/3,245	106/3,158	73/2,689	88/2,735
Department of Education	19/194	19/172	8/158	11/146
Department of Energy	69/786	88/813	30/455	32/365
Department of Health and Human Services	83/618	107/617	115/675	140/727
Department of Homeland Security	74/357	94/350	59/355	69/391
Department of Justice	22/239	24/233	20/231	35/250
Department of Labor	25/151	29/154	28/150	27/158
Department of State	10/83	13/82	7/83	17/94
Department of the Treasury	11/299	19/315	17/286	28/273
Department of Transportation	32/404	38/379	28/356	42/357
Department of Veterans Affairs	5/33	5/32	3/41	6/62
General Services Administration	17/103	26/114	39/111	43/114
Small Business Administration	4/29	4/26	5/27	4/26
Social Security Administration	7/51	7/48	7/53	23/66
Total	567/6,951	624/6,839	520/6,056	663/6,157

Source: GAO analysis of IT Dashboard data as of October 9, 2018. | GAO-19-58.

The Office of Management and Budget (OMB) requires federal agencies to evaluate each investment, or components or systems within the investment, for cloud services, regardless of the overall life cycle stage of the investment.

Agencies are required to report the status of each investment's evaluation as part of the annual budget submission, as noted in OMB's annual capital planning guidance.[69] Table 10 lists the total number of investments using cloud services and the total number of all IT investments for fiscal years 2016 through 2019 for 16 selected agencies, as reported on the IT Dashboard as of October 9, 2018.

APPENDIX IV: SELECTED AGENCY CLOUD SPENDING FOR FISCAL YEARS 2015 THROUGH 2018

The Office of Management and Budget (OMB) requires federal agencies to report total cloud spending based on the cloud deployment model as part of the annual budget submission, as noted in OMB's annual capital planning guidance for fiscal years 2015 through 2018.[70] Table 11 lists the total agency-reported cloud spending and total IT spending for fiscal years 2015 through 2018 for the 16 agencies in our review, as reported on the IT Dashboard as of October 5, 2018.

---

[69] See OMB, *FY 2016 IT Budget–Capital Planning Guidance* (Washington, D.C.: June 27, 2014); *FY 2017 IT Budget–Capital Planning Guidance* (Washington, D.C.: June 16, 2015); *FY 2018 IT Budget–Capital Planning Guidance* (Washington, D.C.: June 30, 2016); *and FY 2019 IT Budget–Capital Planning Guidance* (Washington, D.C.: Aug. 1, 2017).

[70] See OMB, *FY 2015 Guidance on Exhibits 53 and 300* (Washington, D.C.: July 1, 2013); *FY 2016 IT Budget–Capital Planning Guidance* (Washington, D.C.: June 27, 2014); *FY 2017 IT Budget–Capital Planning Guidance* (Washington, D.C.: June 16, 2015); *and FY 2018 IT Budget–Capital Planning Guidance* (Washington, D.C.: June 30, 2016).

Table 11. Cloud and IT spending, as reported on the IT dashboard by 16 selected agencies for fiscal years (FY) 2015-2018 (in millions of dollars)

Agency	IT spending, FY 2015	Cloud spending, FY 2015	IT spending, FY 2016	Cloud spending, FY 2016	IT spending, FY 2017	Cloud spending, FY 2017	IT spending, FY 2018	Cloud spending, FY 2018[a]
Department of Agriculture	3,057.07	80.30	3,418.00	78.53	2,998.08	135.89	2,116.03	55.37
Department of Commerce	2,106.64	222.88	2,312.40	274.93	2,432.70	225.13	2627.07	225.82
Department of Defense	3,0414.27	1,233.65	3,0780.25	1,258.43	17,424.98	153.26	18,585.05	100.21
Department of Education	686.94	62.30	688.56	65.20	699.56	69.17	686.69	6.52
Department of Energy	1,587.75	39.02	1,654.51	45.50	1,818.53	38.33	1,986.30	21.20
Department of Health and Human Services	1,3637.66	64.92	12,566.03	81.41	13,934.64	527.42	5,698.91	150.87
Department of Homeland Security	5,914.24	242.75	6,203.51	230.24	6,767.88	155.20	6,843.29	141.16
Department of Justice	2,646.85	30.92	2,699.14	34.68	2,699.48	34.64	2,788.15	33.56
Department of Labor	667.13	55.68	713.83	60.73	703.38	39.57	715.23	28.68
Department of State	1,576.85	14.92	1,966.18	19.21	1,909.89	80.94	2,286.43	89.95
Department of the Treasury	3,769.68	134.39	3,939.82	145.47	4,226.86	128.45	4,538.30	42.37
Department of Transportation	3,280.70	19.61	3,507.20	25.77	3,105.09	28.18	3,086.04	22.98
Department of Veterans Affairs	4,199.53	29.56	4,403.41	27.91	4,364.56	286.60	4,395.39	280.25
General Services Administration	649.61	20.65	710.16	54.82	768.44	91.46	680.58	81.83
Small Business Administration	102.50	4.32	95.14	4.74	97.90	7.06	101.56	9.01
Social Security Administration	1,861.60	58.14	1,501.32	41.00	1,624.87	38.99	1,618.41	50.30
Total	76,159.01	2,314.01	77,159.44	2,448.56	65,576.82	2,040.26	58,753.44	1,340.08

Source: GAO analysis of agency data on IT Dashboard as of October 5, 2018. | GAO-19-58.

[a]Although agency-reported cloud spending for fiscal year 2018 was obtained from the IT Dashboard on October 5, 2018, the agencies last updated their data between January and March 2018.

APPENDIX V: DESCRIPTION OF CLOUD COMPUTING INVESTMENTS PROVIDED BY SELECTED AGENCIES FOR FISCAL YEAR 2018

Sixteen selected agencies provided us with information on their investments related to spending on cloud services of at least $1 million or more in life-cycle costs.[71] Table 12 identifies the investments for fiscal year 2018 that these agencies submitted to GAO as of October 2018. This list includes the name of the investment, the cloud deployment model, and the cloud service model.

Table 12. Information on investments using cloud services and associated service and deployment model provided by 16 selected agencies for fiscal year 2018, as of October 2018

Agency/Investment	Cloud deployment model	Cloud service model
Department of Agriculture		
Agricultural Marketing Service, Infrastructure Wide Area Network, and De-Militarized Zone, BlackBoard	Public	Software
Agricultural Marketing Service, Infrastructure Wide Area Network, and De-Militarized Zone	Public	Software
Agricultural Marketing Service, Web-Based Supply Chain Management, Business Intelligence, Cloud prototype	Public	Platform
Animal and Plant Health Inspection Service, Electronic Permits System	Private	Platform
Animal and Plant Health Inspection Service, Integrated Plant Health Information System	Private	Platform
Animal and Plant Health Inspection Service, Agricultural Quarantine Activity System	Private	Platform
Animal and Plant Health Inspection Service, Animal Care Information System	Private	Platform

[71] We collected this information to obtain and analyze agency spending and savings on investments using cloud services. We requested that agencies provide spending and savings data broken down by investment, as OMB only required agencies to report total spending by cloud deployment model on the IT Dashboard, and did not require agencies to identify whether any reported savings were cloud-related.

Agency/Investment	Cloud deployment model	Cloud service model
Animal and Plant Health Inspection Service, Certification, Accreditation, Registration, Permitting, and Other Licensing, eFile	Public	Software
Animal and Plant Health Inspection Service, Cloud Bundle	Public	Platform
Animal and Plant Health Inspection Service, Cloud Bundle	Private	Platform
Animal and Plant Health Inspection Service, Enterprise Infrastructure, Cloud	Public	Platform
Animal and Plant Health Inspection Service, Investigation Tracking Enforcement Management System	Private	Platform
Animal and Plant Health Inspection Service, Marketing and Regulatory Programs Business Services Program Support, Cloud	Public	Software
Animal and Plant Health Inspection Service, Marketing and Regulatory Programs Business Services Program Support, Cloud	Public	Software
Animal and Plant Health Inspection Service, Marketing and Regulatory Programs Business Services Program Support, Employee Qualification System 2.0	Public	Platform
Animal and Plant Health Inspection Service, Marketing and Regulatory Programs Business Services Program Support, eSignLive	Public	Software
Animal and Plant Health Inspection Service, Marketing and Regulatory Programs Business Services Program Support, Geographic Information System Cloud	Public	Software
Animal and Plant Health Inspection Service, Phytosanitary Certificate Issuance and Tracking System	Private	Platform
Animal and Plant Health Inspection Service, Phytosanitary Certificate Issuance and Tracking System, Cognos	Private	Platform
Client Technology Services, Infrastructure, Enterprise EMS/OFFICE 365 Product Line	Private	Software
Client Technology Services, Infrastructure, Remedy Cloud	Public	Software
Departmental Management, Office of Human Resources Management, OneUSDA, Onboarding/eRecurit	Private	Software
Departmental Management, Office of Operation, Agriculture Automated Warning and Information Response System, Office of Operation Emergency Notification Systems,	Private	Platform
Departmental Management, Office of Procurement and Property Management, Integrated Acquisition System	Private	Platform
Departmental Management, Office of the Chief Information Officer, Client Technology Services, Enterprise Messaging Systems-Cloud Services, Government Community Cloud migration	Public	Software

Table 12. (Continued)

Agency/Investment	Cloud deployment model	Cloud service model
Economic Research Service, Information Technology Infrastructure General Support System, Cloud Migration	Public	Infrastructure
Enterprise Data Center & Hosting Shared Services	Private	Infrastructure and Platform
Farm Service Agency, 109 Information Technology Management: Bridges to Opportunity	Public	Software
Farm Service Agency, 95 Digital Communications Platforms Support: AskFSA	Public	Software
Farm Service Agency, 95 Digital Communications Platforms Support: GovDelivery	Public	Software
Food Safety and Inspection Service, Web Service, Ask Karen	Public	Platform
Food Safety and Inspection Service, Web Service, AskExpert	Public	Platform
Food Safety and Inspection Service, Web Service, AskFSIS	Public	Platform
Food Safety and Inspection Service, Web Service, CallSat Customer Satisfaction	Public	Platform
Food Safety and Inspection Service, Web Service, FSIS.USDA.Gov site	Public	Platform
Food Safety and Inspection Service, Web Service, GovDelivery Communications Suite	Public	Platform
Food Safety and Inspection Service, Web Service, Small Plant Helpdesk	Public	Platform
Foreign Agricultural Service, Infrastructure, Office Automation, and Telecommunications, Enterprise Data Center	Private	Platform
Foreign Agriculture Service, Global Crop Production Intelligence System	Public	Software
Foreign Agriculture Service, Infrastructure, Office Automation, and Telecommunications, Information Technology Service Management	Public	Software
Foreign Agriculture Service, Infrastructure, Office Automation, and Telecommunications, Internet Mapping Service	Community	Software
Forest Service Computer Base	Public	Software
Forest Service Computer Base: ArcGIS Online	Public	Software
Forest Service Computer Base: Check In/Out	Public	Software
Forest Service Computer Base: ESRI Managed Cloud	Hybrid	Software
Forest Service Computer Base: Pilot	Public	Infrastructure
Forest Service Computer Base: Pinyon	Public	Software

Agency/Investment	Cloud deployment model	Cloud service model
National Agriculture Statistics Service, Information Technology Infrastructure, Information Technology Service Management	Public	Software
National Agriculture Statistics Service, Information Technology Infrastructure, Web Portal, Mobility	Public	Platform
National Resources Conservation Service, Information Technology Management - Conservation Delivery Streamline Initiative	Hybrid	Platform
National Resources Conservation Service, Information Technology Management –Office of the Chief Information Officer, Infrastructure Framework	Public	Infrastructure
National Resources Conservation Service, Information Technology Management –Office of the Chief Information Officer, Infrastructure Framework	Public	Infrastructure
National Resources Conservation Service, Information Technology Management, Box	Public	Software
National Resources Conservation Service, Information Technology Management, Project Portfolio Management Online	Public	Software
Office of the Chief Financial Officer, Federal Shared Service Provider, Pegasys, Cloud Migration	Private	Infrastructure
Office of the Chief Financial Officer, Financial Management Modernization Initiative, Cloud Migration	Private	Platform
Risk Management Agency, 13 Enterprise Information Technology Acquisition, Escrow	Community	Platform
Department of Commerce		
Bureau of Economic Analysis Regional Dashboard	Public	Platform
Bureau of Economic Analysis website	Public	Platform
Bureau of Industry and Security, Collaboration Tools	Public	Platform and Software
Bureau of Industry and Security, Infrastructure	Public	Infrastructure
Census, Cloud Computing Services	Hybrid	Infrastructure
National Oceanic and Atmospheric Administration, Collaboration Tools	Community	Software
National Oceanic and Atmospheric Administration, Enterprise Service Management, Workflow Automation	Community	Software
National Oceanic and Atmospheric Administration, Web Operations Center	Hybrid	Platform
Office of the Secretary, Office of Acquisition Management, Commerce Business Environment	Private	Infrastructure
Office of the Secretary, Office of Financial Management, Commerce Business Systems	Private	Infrastructure

Table 12. (Continued)

Agency/Investment	Cloud deployment model	Cloud service model
Office of the Secretary, Office of Financial Management, Sunflower Enterprise Personal Property Management System	Private	Infrastructure
Office of the Secretary, Office of Human Resource Management, Commerce Learning Center	Private	Software
Office of the Secretary, Office of Human Resource Management, General Support Systems	Private	Infrastructure
Office of the Secretary, Office of Human Resource Management, Human Resource Management System	Private	Software
Office of the Secretary, Office of Human Resource Management, NFC Fee for Service	Private	Software
Office of the Secretary, Office of Human Resource Management, Web-based Time and Attendance System	Private	Infrastructure
Department of Defense		
Academic and Professionally Accredited Enterprise Education Enclave	Public	Software
Acquisition Career Management System	Public	Infrastructure
Advanced Distributed Learning Service	Public	Platform
Air Force Combined Mishap Reduction System	Public	Platform
Aircraft Parts Tracking	Public	Platform
All Other Navy/United States Marine Corps Computing Infrastructure	Private	Platform
All Weapons Information System	Public	Infrastructure
Alternative Dispute Resolution System	Public	Software
Case Management Tracking Analysis & Reporting System	Hybrid	Platform
Cellular Phones	Private	Platform
Chief of Information Social Media Audio Visual Services	Public	Infrastructure
Civilian Personnel Online	Hybrid	Infrastructure
Classified - Various	Private	Platform
Classified - Various	Private	Software
Classified - Various	Community	Infrastructure
Continuous Monitoring Program	Private	Platform
Contracting Information Technology	Private	Infrastructure
Core Data Centers and Installation Processing Nodes	Private	Infrastructure
Core Data Centers and Installation Processing Nodes	Private	Platform
Core Services Architecture	Private	Infrastructure
Customer Relationship Management Software as a Service	Private	Software
Data Collection And Scheduling Tool	Community	Platform
Defense Contract Audit Agency Integrated Information Network	Public	Software

Agency/Investment	**Cloud deployment model**	**Cloud service model**
Defense Contract Audit Agency LiveLink Records Management System	Public	Infrastructure
Defense Contract Audit Agency LiveLink Records Management System	Public	Software
Defense Environmental Network and Information Exchange-Knowledge-Based Corporate Reporting System	Public	Infrastructure
Defense Talent Management System	Private	Software
Department of Defense Electronic Official Personnel Folder	Private	Software
Department of the Army Mobilization Processing System	Private	Infrastructure
Department of the Army Mobilization Processing System - Unclassified	Private	Infrastructure
Department of the Navy Heritage Asset Management System - Collections Management System	Private	Infrastructure
Department of the Navy Heritage Asset Management System - Collections Management System	Private	Infrastructure
Electronic 332 System	Private	Platform
Electronic Construction Management System	Private	Infrastructure
Enterprise Information Environment Computing Infrastructure Assets	Public	Infrastructure
Enterprise Information Environment Core Enterprise Services	Public	Infrastructure
Enterprise Information Environment Core Enterprise Services	Public	Software
Enterprise Information Environment Mission Area Computing Infrastructure Domain/Computing Services_Server Services	Private	Platform
Enterprise Information Environment Mission Area Core Enterprise Services/App Services_Service Desk	Private	Platform
Enterprise Proactive Real-Property Interactive Space Management System	Private	Infrastructure
Enterprise Protection Risk Management	Community	Infrastructure
Family Separation Allowance Information Management Capabilities	Hybrid	Infrastructure
Family Separation Allowance Information Management Capabilities	Hybrid	Infrastructure
Fleet Home Town News Center	Public	Infrastructure
Fleet Home Town News Center	Public	Platform
Force Management Tools	Public	Infrastructure
Fourth Estate Manpower Tracking System	Private	Platform
Global Energy Information System	Public	Platform
Information Technology Division SharePoint, Web Services	Public	Infrastructure
Integrated Budget Documentation & Execution System	Public	Platform

Table 12. (Continued)

Agency/Investment	Cloud deployment model	Cloud service model
Integrated Digital Environment Service Center	Private	Software
Joint Training Capability Development	Private	Infrastructure
Keyport Integrated Production and Application Development System	Community	Infrastructure
Logistics Execution Information System	Private	Platform
Maintenance & Ship Work Planning	Private	Infrastructure
Maintenance work request	Public	Platform
Manpower Authorization Man-day Management System	Public	Infrastructure
Manpower Programming and Execution System	Public	Infrastructure
Manpower, Personnel, Training and Education Network, Operations, and Other Communications Support	Public	Software
Marine Corps Recruiting Command Advertising Information Technology System	Public	Infrastructure
Mobilization Common Operating Picture	Private	Infrastructure
Mobilization Common Operating Picture Messaging System	Private	Infrastructure
Mobilization Common Operating Picture - Unclassified	Private	Infrastructure
Mobilization Deployment Information System	Private	Infrastructure
Module Test Repair Activity	Community	Platform
Naval Audit Service Information Management System	Private	Platform
Naval Reserve Force Reserve Affairs Capabilities	Hybrid	Software
Naval Systems Engineering Resource Center - Unclassified	Hybrid	Infrastructure
Naval Tactical Command Support System	Private	Platform
Naval War College	Public	Software
Naval Sea Systems Command Headquarters Next Generation Enterprise Network	Private	Platform
Naval Sea Systems Command Headquarters Related Technical Activities	Public	Infrastructure
Navy Chief Of Information Website	Public	Infrastructure
Navy Consolidated Brig Physical Security System	Public	Software
Navy Maritime Maintenance Enterprise Solution	Community	Infrastructure
Network Enterprise Center Staff Operations Costs	Private	Infrastructure
Next Generation Enterprise Network Increment 1	Private	Platform
NLign Analytics	Private	Platform
Ogden Air Logistics Complex's 309th Software Maintenance Group Work Center	Public	Platform
Overseas Contingency Operations Individual	Private	Infrastructure
Overseas Contingency Operations Temporary Change of Station	Private	Infrastructure
Pacific Fleet Base-Level Communications Capabilities	Private	Platform

Agency/Investment	Cloud deployment model	Cloud service model
Pacific Fleet Systems Support	Private	Platform
Part Number Supply Support Request System	Private	Platform
Physical Sciences Laboratory Information Management System Plus	Private	Platform
Planned Labor Workload Tool	Private	Platform
Priority Requisition Information Management Enterprise	Private	Platform
Program Manager Soldier Weapons Collaborative Data Environment	Private	Platform
Programmed Depot Maintenance Scheduling System Web	Public	Platform
Promotion Testing System	Private	Platform
Propulsion Engineering Workspace	Private	Platform
Science & Technology	Private	Infrastructure
Science & Technology	Private	Software
Senior Leader Development Management System	Private	Infrastructure
Shipyard Management Information System Support	Private	Platform
Shipyard Management Information System Support	Public	Platform
Space and Naval Warfare Systems Command System Center Information Technology Division Data Center	Private	Platform
Strategic Management System	Private	Infrastructure
System Metric and Reporting Tool	Private	Platform
System Metric and Reporting Tool	Private	Software
Technical Assurance System	Community	Infrastructure
Time Labor Management System	Public	Software
Tools for Environmental Management Configuration Information	Public	Platform
Tools for Environmental Management Restoration Information	Public	Platform
Total Ammunition Management Information System	Public	Infrastructure
Total Ammunition Management Information System	Public	Infrastructure
Total Component Management	Public	Platform
Tour Of Duty	Private	Infrastructure
Training Scheduling System	Private	Platform
Transportation Coordinators' Automated Information for Movements System II	Private	Software
Warfighting Battlespace Awareness Tools	Hybrid	Infrastructure
Warfighting Command and Control Tools	Public	Infrastructure
Warfighting Mission Area Force Support, Force Management	Public	Platform
Web Compliance Assessment and Sustainment System	Community	Platform
Yellow Ribbon Reintegration Program Web Portal	Private	Infrastructure
Department of Education		
Cloud Based Storage	Community	Software

Table 12. (Continued)

Agency/Investment	Cloud deployment model	Cloud service model
Federal Student Aid Information Center	Community	Software
Institute of Education Sciences Support (consolidated investment)	Public	Infrastructure
Next Generation Data Center	Private	Infrastructure
Ombudsman Case Tracking System, Federal Student Aid Feedback and Dispute System	Community	Software
Department of Energy		
Cloud Service Provider	Community	Software
Cloud Service Provider - Analysis	Public	Infrastructure
Collaboration: Video and Voice Services	Public	Software
Collaboration: Video and Voice Services – Oak Ridge National Laboratory	Public	Software
Co-Located Data Center	Hybrid	Infrastructure
Content Management System	Community	Software
Content Management System	Hybrid	Platform
Content Management System Vendor 1	Public	Infrastructure
Content Management System Vendor 2	Community	Infrastructure
Digital Experience Platform	Public	Software
Digital Experience Platform - Community Cloud	Community	Software
Digital Experience Platform - Community Infrastructure as a Service	Community	Infrastructure
Digital Experience Platform - Electronic Infrastructure	Public	Software
Digital Experience Platform - E-Mail Software as a Service Infrastructure	Public	Software
Digital Experience Platform - Enhanced Licenses	Community	Software
Digital Experience Platform - Government Community Cloud	Community	Software
Digital Experience Platform - Implementation	Community	Software
Digital Experience Platform - Licenses	Public	Software
Digital Experience Platform - Not Mission Critical	Community	Software
Digital Experience Platform - Private Cloud	Private	Software
Digital Experience Platform - Public Cloud E-Mail	Public	Software
Digital Experience Platform – Enterprise License Agreement National Renewable Energy Laboratory	Community	Software
Environmental, Health and Safety - Public Software as a Service Cloud	Public	Software
Finance and Human Resources	Private	Software
Finance and Human Resources	Private	Software
Finance and Human Resources - Cloud Labor	Public	Software

Agency/Investment	**Cloud deployment model**	**Cloud service model**
Finance and Human Resources - Infrastructure	Public	Software
Finance and Human Resources	Public	Software
Finance and Human Resources - Public Software as a Service Cloud Infrastructure	Public	Software
Industrial Plant Management	Community	Platform
Information Technology Service Manager - Licenses	Public	Software
Information Technology Service Manager – Software as a Service Infrastructure	Public	Software
Information Technology Service Manager - Workflow	Private	Platform
Information Technology Service Manager - Workflow Licenses	Private	Software
Information Technology Service Manager - Workflow Oak Ridge National Laboratory	Community	Platform
Investment Modeling	Community	Infrastructure
Online Training	Community	Software
Security Software as a Service	Hybrid	Software
Survey Tool	Community	Software
Training	Community	Software
Department of Health and Human Services		
Administration for Community Living Digital	Hybrid	Infrastructure, Platform and Software
Agency for Healthcare Research and Quality Cloud	Public	Infrastructure
Centers for Disease Control and Prevention, End User Infrastructure	Private	Infrastructure and Platform
Centers for Disease Control and Prevention, Information Technology Infrastructure	Private	Infrastructure and Platform
Centers for Disease Control and Prevention, Public Health Information Network: BioSense	Public	Infrastructure and Platform
Centers for Medicare and Medicaid Services, Accountable Care Organization, Model Support	Hybrid	Infrastructure
Centers for Medicare and Medicaid Services, Beneficiary e-Services	Public	Infrastructure, Platform and Software
Centers for Medicare and Medicaid Services, Case Hearings and Appeals	Public	Infrastructure, Platform and Software
Centers for Medicare and Medicaid Services, Enterprise Identity Management	Hybrid	Infrastructure, Platform and Software
Centers for Medicare and Medicaid Services, Healthcare Web Support	Private	Infrastructure, Platform and Software

Table 12. (Continued)

Agency/Investment	Cloud deployment model	Cloud service model
Centers for Medicare and Medicaid Services, Innovation Core Systems	Private	Infrastructure and Platform
Centers for Medicare and Medicaid Services, IT Infrastructure	Hybrid	Infrastructure and Platform
Centers for Medicare and Medicaid Services, Medicaid & Children's Health Insurance Program, Business Information and Solutions	Private	Infrastructure and Platform
Centers for Medicare and Medicaid Services, Oncology Payment Model	Hybrid	Infrastructure
Centers for Medicare and Medicaid Services, Plan Enrollment, Health Plan Management System	Private	Infrastructure and Platform
Centers for Medicare and Medicaid Services, Quality Payment Program	Public	Infrastructure and Platform
Continued Implementation of Homeland Security Presidential Directive-12 (formerly known as IAM@HHS)	Private	Infrastructure and Platform
Food and Drug Administration, Center for Devices and Radiological Health Administrative Support Systems	Community	Infrastructure and Platform
Food and Drug Administration, Center for Devices and Radiological Health High Performance Computing	Private	Infrastructure and Platform
Food and Drug Administration, Office of Information Management and Technology Email	Hybrid	Infrastructure and Platform
Food and Drug Administration, Office of Information Management and Technology, Information Technology Infrastructure & Data Center Management	Hybrid	Infrastructure and Platform
Department of Health and Human Services, Cybersecurity Operations Program Investment	Public	Infrastructure and Platform
Department of Health and Human Services, Email as a Service Investment	Community	Software
Department of Health and Human Services, Enterprise Human Capital Management Investment	Hybrid	Infrastructure and Platform
Department of Health and Human Services, Financial Business Intelligence System Investment	Private	Infrastructure and Platform
Department of Health and Human Service, Grant Solutions Center of Excellence	Public	Infrastructure and Platform
Department of Health and Human Services, Unified Financial Management System Investment	Private	Infrastructure and Platform
Health Resources and Services Administration, Bureau of Health Workforce, National Practitioner Data Bank	Public	Infrastructure and Platform
Health Resources and Services Administration, Office of Information Technology Core Business Support System	Public	Software

Agency/Investment	**Cloud deployment model**	**Cloud service model**
Indian Health Service, Health Information Technology Systems and Support	Private	Infrastructure, Platform and Software
Indian Health Service, Infrastructure, Office Automation, & Telecommunications	Private	Infrastructure
National Institutes of Health, Electronic System Electronic Research Administration	Hybrid	Infrastructure and Platform
National Institutes of Health, Electronic System Business System	Private	Infrastructure and Platform
National Institutes of Health Information, Technology Data Centers and Cloud Services	Hybrid	Infrastructure, Platform and Software
National Institutes of Health Information, Technology End User Devices, Tools & Support	Hybrid	Infrastructure, Platform and Software
National Institutes of Health, National Cancer Institute, Administration and Management Support Applications and Tools	Public	Software
National Institutes of Health, National Cancer Institute, Enterprise Architecture, Capital Planning, Information Technology Management & Oversight	Public	Software
National Institutes of Health, National Cancer Institute, Genomic Data Commons	Private	Infrastructure and Platform
National Institutes of Health, National Cancer Institute, Sci Mission Clinical Trials and Management Support Tools	Private	Software
National Institutes of Health, National Cancer Institute, Sci Mission Research Knowledge Dissemination Tools	Hybrid	Infrastructure, Platform and Software
National Institutes of Health, National Heart, Lung, and Blood Institute, Administration and Management Support Applications and Tools	Public	Infrastructure and Platform
National Institutes of Health, National Heart, Lung, and Blood Institute, Sci Mission Biomedical Data Repositories & Analytical Tools	Private	Software
National Institutes of Health, National Heart, Lung, and Blood Institute, Sci Mission Extramural Research Management and Support Tools	Public	Software
National Institutes of Health, National Institute of Allergy and Infectious Diseases, Sci Mission Biomedical Data Repositories & Analytical Tools	Public	Infrastructure and Platform
National Institutes of Health, National Institute of Allergy and Infectious Diseases, Sci Mission Extramural Research Management and Support Tools	Public	Infrastructure and Platform

Table 12. (Continued)

Agency/Investment	Cloud deployment model	Cloud service model
National Institutes of Health, National Institute of Child Health and Human Development, Sci Mission Biomedical Data Repositories & Analytical Tools	Public	Infrastructure and Platform
National Institutes of Health, National Institute of Mental Health, Sci Mission Biomedical Data Repositories & Analytical Tools	Public	Infrastructure and Platform
Office of Inspector General, Data Analytics	Hybrid	Infrastructure, Platform and Software
Office of Inspector General, Support Services	Hybrid	Infrastructure, Platform and Software
Office of the Secretary, Assistant Secretary for Administration, E-Gov Travel Support Center of Excellence	Public	Infrastructure and Platform
Office of the Secretary, Assistant Secretary for Administration, Health and Human Services Asset - Property Management Information System (Sunflower)	Private	Infrastructure and Platform
Office of the Secretary, Assistant Secretary for Administration, Health and Human Services Careers System	Community	Infrastructure and Platform
Office of the Secretary, Assistant Secretary for Administration Learning Management System	Public	Infrastructure and Platform
Office of the Secretary, Assistant Secretary for Administration Physical Access Control Systems' Affiliation Project	Private	Infrastructure and Platform
Office of the Secretary, Assistant Secretary for Administration, PSC Revenue, Invoicing and Cost Estimation System	Private	Infrastructure and Platform
Office of the Secretary, Office of the Assistant Secretary for Financial Resources, Departmental Contracts Information System	Private	Infrastructure and Platform
Office of the Secretary, Office of the Assistant Secretary for Financial Resources, Department of Health and Human Services Consolidated Financial Reporting System	Private	Infrastructure and Platform
Office of the Secretary, Office of the Assistant Secretary for Health, Commissioned Corps Force Management Solution	Private	Infrastructure and Platform
Office of the Secretary, Office of the Assistant Secretary for Health, Office on Women's Health Websites Communications Initiative	Private	Infrastructure and Platform

Agency/Investment	**Cloud deployment model**	**Cloud service model**
Office of the Secretary, Office of the Assistant Secretary for Public Affairs, Department of Health and Human Services Web Management Investment	Private	Infrastructure and Platform
Office of the Secretary, Office of Medicare Hearings and Appeals, Electronic Case Adjudication Processing Environment	Private	Infrastructure and Platform
Department of Homeland Security		
Coast Guard, One View-Ozone Widget Framework	Public	Infrastructure
Department of Homeland Security, Management Directorate, Enterprise Data Management Office, Collibra Data Governance	Public	Software
Department of Homeland Security, Management Directorate, Enterprise Information Technology Service Division, Workplace as a Service	Private	Software
Department of Homeland Security, Management Directorate, Enterprise IT Service Division, Public Cloud Web Content Management as a Service	Public	Infrastructure
Department of Homeland Security, Management Directorate, Infrastructure Operations, Office 365 Software	Public	Software
Department of Homeland Security, Management Directorate, Office of the Chief Information Officer, OneNet, Email	Private	Software
Department of Homeland Security, Management Directorate, Office of the Chief Information Officer, Homeland Security Information Network	Public	Infrastructure
Domestic Nuclear Detection Office, Joint Analysis Center Collaborative Information System 2.0	Public	Infrastructure
Federal Emergency Management Agency, Applicant Case Tracker	Public	Infrastructure
Federal Emergency Management Agency, Crisis Management System	Public	Infrastructure
Federal Emergency Management Agency, Deployment Tracking System	Public	Infrastructure
Federal Emergency Management Agency, HSEEP System, Preparedness Toolkit	Public	Infrastructure
Federal Emergency Management Agency, Human Capital Investments, Career Path Tool	Public	Infrastructure
Federal Emergency Management Agency, Infrastructure, Email as a Service	Public	Software
Federal Emergency Management Agency, Internet, Digital Imagery and Media on Demand,	Public	Infrastructure

Table 12. (Continued)

Agency/Investment	Cloud deployment model	Cloud service model
Federal Emergency Management Agency, National Flood Insurance Program, Community Information System, Citizen Corps	Public	Software
Federal Emergency Management Agency, National Flood Insurance Program, Information Technology Systems and Services, ReInsurance Broker System	Public	Infrastructure
Federal Emergency Management Agency, Recovery Management Tools, Disaster Management Support Environment Cloud Environment	Public	Infrastructure
Federal Law Enforcement Training Centers, All Other IT Costs, Mobile Application	Public	Infrastructure
Federal Law Enforcement Training Centers, Infrastructure, Enterprise Training Cloud	Public	Infrastructure
Federal Law Enforcement Training Centers, Online Campus, eFLETC	Public	Software
National Protection and Programs, Directorate Automated Biometric Identification System, Operational Data Store	Public	Infrastructure
National Protection and Programs, Directorate Interagency Security Committee-Compliance System	Public	Infrastructure
National Protection and Programs, Directorate National Cybersecurity and Communications Integration Center, Trusted Automated eXchange of Indicator Information Server	Public	Infrastructure
National Protection and Programs Directorate, Communication Assets Survey and Mapping tool, Next Generation	Public	Infrastructure
Office of Inspector General, Infrastructure, Data Analytics Cloud System	Public	Infrastructure
Transportation Security Administration, Information Technology Infrastructure Program, Alert & Warning System	Public	Software
Transportation Security Administration, Information Technology Infrastructure Program, Core Services	Public	Infrastructure
Transportation Security Administration, Information Technology Infrastructure Program, DEV	Public	Software
Transportation Security Administration, Information Technology Infrastructure Program, Enterprise App Central	Public	Infrastructure
Transportation Security Administration Information Technology Infrastructure Program, Security Platform for Enterprise Compliance, Testing, and Reporting,	Public	Infrastructure

Agency/Investment	Cloud deployment model	Cloud service model
U.S. Citizenship and Immigration Services Information Security, Identity Credential and Access Management	Public	Infrastructure
U.S. Citizenship and Immigration Services MyUSCIS	Public	Infrastructure
U.S. Citizenship and Immigration Services Scheduling, National Appointment Scheduling System	Public	Infrastructure
U.S. Citizenship and Immigration Services Standard Tool Program, Salesforce Tracking Activities and Relationships System	Public	Software
U.S. Citizenship and Immigration Services Standard Tool Program, Standard Development Tool	Public	Software
U.S. Citizenship and Immigration Services Transformation, Electronic Immigration System 2	Public	Infrastructure
U.S. Coast Guard Incident Management Software System, Incident Management Software System	Public	Infrastructure
U.S. Customs and Border Patrol Automated Commercial Environment	Public	Infrastructure
U.S. Customs and Border Patrol Biometric Entry-Exit, Traveler Verification Service	Public	Infrastructure
U.S. Customs and Border Patrol Electronic System for Travel Authorization, Data Cube	Public	Infrastructure
U.S. Customs and Border Patrol Electronic Visa Update System	Public	Software
U.S. Customs and Border Patrol Office 365, Small OIT Projects	Public	Software
U.S. Customs and Border Patrol Right Now Web for Customer Service Center, Complaint Management System,	Public	Infrastructure
U.S. Immigration and Customs Enforcement Analytics, FALCON	Public	Infrastructure
U.S. Immigration and Customs Enforcement Critical Foundational Infrastructure, Cloud Collaboration Software Suite	Public	Software
U.S. Immigration and Customs Enforcement Critical Foundational Infrastructure, Cloud General Support System	Public	Infrastructure
U.S. Immigration and Customs Enforcement Executive Secretary Systems, Correspondence and Task Tracking System	Public	Infrastructure
U.S. Immigration and Customs Enforcement Investigative Application Support, HSI Application Platform	Public	Infrastructure
U.S. Immigration and Customs Enforcement Office of the Principal Legal Advisor Systems, Electronic Service	Public	Infrastructure
U.S. Immigration and Customs Enforcement Student and Exchange Visitor Information System Modernization	Public	Infrastructure
Department of Justice		

Table 12. (Continued)

Agency/Investment	Cloud deployment model	Cloud service model
Bureau of Alcohol, Tobacco, Firearms & Explosives Cloud Computing	Community	Infrastructure, Platform and Software
Bureau of Alcohol, Tobacco, Firearms & Explosives Mission and Business Systems Support	Community	Infrastructure
Bureau of Prisons Network	Community	Infrastructure and Platform
Drug Enforcement Agency Firebird	Private	Infrastructure
Drug Enforcement Agency Speedway	Private	Infrastructure
Federal Bureau of Investigation Criminal Justice Information Services Disaster Recovery	Private	Infrastructure
Federal Bureau of Investigation Combined DNA Index System	Private	Infrastructure and Platform
Federal Bureau of Investigation Digital Collection	Private	Infrastructure, Platform and Software
Federal Bureau of Investigation Human Resource Source	Private	Infrastructure and Platform
Federal Bureau of Investigation Intelligence Community Information Technology Enterprise	Public	Software
Federal Bureau of Investigation Innocence Lost Database Web Archival Tool	Private	Infrastructure and Platform
Federal Bureau of Investigation Law Enforcement Online	Private	Software
Federal Bureau of Investigation Mobility Program Office	Public	Infrastructure and Platform
Federal Bureau of Investigation National Data Exchange Program	Private	Infrastructure and Platform
Federal Bureau of Investigation National Cyber Investigative Joint Task Force Lighthouse	Private	Infrastructure and Platform
Federal Bureau of Investigation Prometheus	Private	Platform
Justice Management Division Cybersecurity Initiatives	Public	Infrastructure and Platform
Justice Management Division Data Centers Initiatives	Community	Infrastructure and Platform
Justice Management Division Information Technology Planning and Management	Public	Software
Justice Management Division Information Technology Transformation Initiatives	Hybrid	Infrastructure and Platform
Office of Justice Programs Infrastructure and Information Technology Management	Public	Infrastructure and Platform

Agency/Investment	Cloud deployment model	Cloud service model
Office of Justice Programs Servers and Storage	Community	Infrastructure
United States Attorneys Information Technology Program Management	Public	Infrastructure and Platform
United States Attorneys Information Technology Security Program	Public	Infrastructure and Platform
United States Attorneys Litigation Technology Service Center	Public	Infrastructure
United States National Central Bureau Federal Local Information Sharing	Community	Software
Justice Management Division Cybersecurity Services, Justice Security Operations Center	Public	Software
United States Attorneys Office Automation (Non-Justice Consolidated Office Network)	Private	Infrastructure, Platform and Software
United States Marshal Service Mission Modernization, Capture	Community	Infrastructure and Platform
Department of Labor		
Cloud Email	Public	Software
Department of Labor Enterprise Cloud	Hybrid	Infrastructure, Platform and Software
Department of Labor Case Management Platform	Hybrid	Platform
Department of State		
Application Services	Private	Software
Architecture Services	Public	Software
Architecture Services	Hybrid	Platform
Buildings Management Integrated Systems	Private	Platform
Centralized Data Collection and Integration System	Public	Software
Centralized Data Collection and Integration System	Hybrid	Platform
Data Center Services and Hosting	Public	Platform
Data Center Services and Hosting	Private	Software
Defense Trade Services	Hybrid	Platform
Defense Trade Services	Private	Software
Enterprise Network and Bandwidth Services	Private	Software
Enterprise Network and Bandwidth Services	Hybrid	Platform
Global E-Travel Program	Community	Software
Information Resource Management System, Operations and Maintenance	Public	Software
Information Resource Management System, Operations and Maintenance	Public	Platform
myServices	Private	Software
Department of the Treasury		

Table 12. (Continued)

Agency/Investment	Cloud deployment model	Cloud service model
Bureau of Engraving and Printing, Manufacturing Support Suite, Maximo	Private	Software
Cloud-based e-mail and collaboration and productivity tools	Hybrid	Software
Community Development Financial Institutions Fund, Awards Management Information System	Community	Software
Workplace.gov Community Cloud Infrastructure	Community	Infrastructure, Platform and Software
Fiscal Agency Accounting Services investment, Business Suite	Private	Infrastructure, Platform and Software
Fiscal Central Accounting Services investment	Private	Infrastructure and Platform
Fiscal Enterprise Initiatives and Capabilities investment, Financial Information Repository	Public	Infrastructure and Platform
Fiscal Information Technology Infrastructure Services investment, Messaging and Collaboration	Public	Infrastructure, Platform and Software
Fiscal USAspending.gov investment, Data Act Operating Infrastructure	Community	Infrastructure and Platform
Fiscal USAspending.gov	Community	Infrastructure and Platform
IRS.gov, Integrated Enterprise Portal	Public	Infrastructure
Office of the Comptroller of the Currency, Information Technology Customer Support	Community	Software
Office of the Comptroller of the Currency, Mobile Data Management	Community	Platform
Office of the Comptroller of the Currency, email	Community	Software
Tax and Trade Bureau, General Support System, Mission applications Support Services	Private	Infrastructure
Tax and Trade Bureau, Information Technology Infrastructure Mainframes and Servers Services and Support	Private	Infrastructure
Tax and Trade Bureau, Information Technology, Infrastructure End User Systems and Support	Private	Infrastructure and Software
Department of Transportation		
Department of Transportation: Common Operating Environment	Public	Infrastructure and Platform
Department of Transportation: Common Operating Environment	Public	Software

Agency/Investment	Cloud deployment model	Cloud service model
Department of Transportation: Common Operating Environment	Public	Platform
Federal Aviation AdministrationXX137: Enterprise IT Management	Private	Platform and Software
Federal Aviation AdministrationXX141: Infrastructure/Operations Management	Hybrid	Infrastructure and Platform
Federal Aviation AdministrationXX172: Emergency Notification System Program	Hybrid	Software
Federal Highway AdministrationX040: Information Technology Enterprise Services (formerly consolidated with DOTXX070)	Public	Software
Federal Motor Carrier Safety Administration201: Legacy Inspection System	Public	Infrastructure
Federal Motor Carrier Safety Administration603: Hosting Data Center	Public	Infrastructure
Federal Motor Carrier Safety Administration704: Customer Relationship Management System	Public	Software
Federal Transit AdministartionXX028: Information Management Platform	Community	Platform
National Highway Transit Safety Administration348: Grants Management Suite	Public	Software
National Highway Transit Safety Administration020: Artemis	Public	Software
Department of Veterans Affairs		
4Cast Financial Planning	Hybrid	Platform
Appeals Modernization	Hybrid	Infrastructure
Beneficiary Travel Self-Service System	Hybrid	Infrastructure
Captivate Prime	Hybrid	Software
Centralized Adjudication and Background Investigation System	Hybrid	Infrastructure
Clinician User Interface	Hybrid	Infrastructure
Community Care Referrals and Authorization	Private	Software
Data Access Services	Hybrid	Infrastructure
Digital Veterans Platform	Hybrid	Infrastructure
Email as a Service	Hybrid	Software
Enterprise Acquisition Systems	Hybrid	Platform
Enterprise Services Collaboration Portal	Hybrid	Platform
Environment Of Care	Hybrid	Infrastructure
Financial Services Center Informatica	Hybrid	Infrastructure
Human Resources, Personnel and Accounting System, Personnel and Accounting Integrated Data Decommission	Hybrid	Infrastructure
Identity Access Management- Enterprise	Hybrid	Infrastructure

Table 12. (Continued)

Agency/Investment	Cloud deployment model	Cloud service model
Information Technology Operations, Customer Experience Environment	Hybrid	Infrastructure
Information Technology Virtual Campus	Hybrid	Infrastructure
Insurance Capture Buffer Web	Hybrid	Infrastructure
Integrated Research Information System	Hybrid	Infrastructure
Loan Electronic Reporting Interface Redesign	Hybrid	Infrastructure
Loan Guaranty	Hybrid	Infrastructure
Medical Appointment Scheduling System	Hybrid	Infrastructure
Medical Care Collections Fund, Electronic Data Interchange, Transaction Applications Suite	Hybrid	Infrastructure
Mobile Activity Tracking Research Project	Hybrid	Infrastructure
Mobile Application Program - VA Online Scheduling	Hybrid	Infrastructure
Mobile Performance and Operational Web-Enabled Reports	Hybrid	Software
Online Patient Self-scheduling System	Hybrid	Infrastructure
Office of Small and Disadvantaged Business Utilization Vet Biz Portal	Hybrid	Infrastructure
Provider Profile Management System	Hybrid	Infrastructure
Pulse	Hybrid	Platform
Recruit, Enroll, Engage, Follow-up	Hybrid	Infrastructure
Remote Procedure Call Emulation – VistA Adaptive Maintenance	Hybrid	Infrastructure
Vet Ride	Commercial	Software
Veteran Oriented Interactive Customer Evaluation	Private	Infrastructure and Software
Veterans360	Hybrid	Infrastructure
Veterans Benefits Management System and Agile Integrated Development Environment	Hybrid	Infrastructure
Veterans Health Administration, Future Technologies Lab	Hybrid	Infrastructure and Software
Vets.gov Website	Hybrid	Infrastructure
VistA Security Remediation	Hybrid	Platform
VistA Security Scanning	Hybrid	Platform
VistA Standardization and Virtualization	Hybrid	Platform
General Services Administration		
Analytics Service Platform	Public	Infrastructure
Cloud-Based Email and Collaboration Platform	Hybrid	Software
Cloud-Based Service Delivery Platform	Community	Platform
Cloud.gov	Public	Platform
Cloud Infrastructure	Private	Infrastructure

Agency/Investment	Cloud deployment model	Cloud service model
Cloud Management Platform Website hosting	Public	Infrastructure
Common Gateway Interfact Website Hosting	Community	Infrastructure
eGov Innovation	Public	Infrastructure
Electronic Project Management	Public	Software
eSignLive, E-Signature and Digital Signature Service	Hybrid	Software
Frameworks	Private	Software
Human Capital Information Technology Systems	Public	Infrastructure and Software
Integrated Award Environment	Hybrid	Software
Login.gov (Consumer Identity Platform)	Public	Infrastructure and Software
National Computerized Maintenance Management System	Community	Software
Suite of back office and Customer Relationship Management	Community	Platform and Software
Small Business Administration		
Certify.sba.gov	Private	Platform
Disaster Credit Management Modernization	Hybrid	Infrastructure and Platform
SBA.gov	Public	Platform
Small Business Administration Cloud	Hybrid	Infrastructure
Social Security Administration		
Anti-Fraud Enterprise Solution	Public	Infrastructure
Disability Case Processing System	Public	Infrastructure and Platform
Enterprise Data Warehouse	Public	Infrastructure

Source: GAO analysis of agency-provided cloud computing data. | GAO-19-58.

Appendix VI: Profiles of Selected Cloud Computing Acquisitions

The following nine cloud investment profiles illustrate the variety of benefits that the selected agencies in our review had realized from the acquisition of cloud services.[72] These profiles describe the cloud investment, costs, key benefits, and the savings or avoidances associated with implementation of cloud services. In addition, the profiles detail how,

[72] Additional detail on how these agencies were selected can be found in appendix I.

among other things, the acquisition of cloud services enabled the agency to overcome previous challenges with legacy systems and acquire more cost-effective, efficient, and responsive IT resources in order to meet mission needs.

Treasury's Cloud Acquisition Supports Enterprise Resource Planning

Each year, the Department of the Treasury's (Treasury) Bureau of Engraving and Printing prints billions of dollars—referred to as Federal Reserve notes—for delivery to the Federal Reserve System. According to the Chief of the Bureau of Engraving and Printing's Office of Enterprise Solutions, in October 2012, the bureau's CIO deployed an enterprise resource planning system to the cloud to improve efficiency and operations, and enhance the availability, security, and performance of its systems that manage the daily business activities of the bureau.

Previously, the bureau had used 16 separate IT legacy systems that were facing technological obsolescence and required heavy customization using old programming languages. Officials in the bureau reported that they spent time updating and maintaining the hardware and software for these systems to minimize downtime that led to the need for staff to work overtime and decreased customer satisfaction with the system. In addition, the legacy systems were not integrated and certain tasks were performed manually, including data entry, aggregation, and product quality checks. Further, the legacy systems lacked robust data validation features; thus, database administrators had to spend time making corrections to data submitted by users.

By moving to the cloud, the bureau replaced the 16 systems with software as a service that required no customization. As a result, officials in the bureau reported that the bureau was able to significantly improve its operations and decision making. Further, officials stated that the bureau's use of software as a service enabled it to procure cloud services which helped to ensure that the new system had increased availability, performance, and security to prevent delays and downtime in operations.

The bureau also implemented features to improve data entry to reduce user errors. Figure 3 provides a summary of Treasury's cloud acquisition.

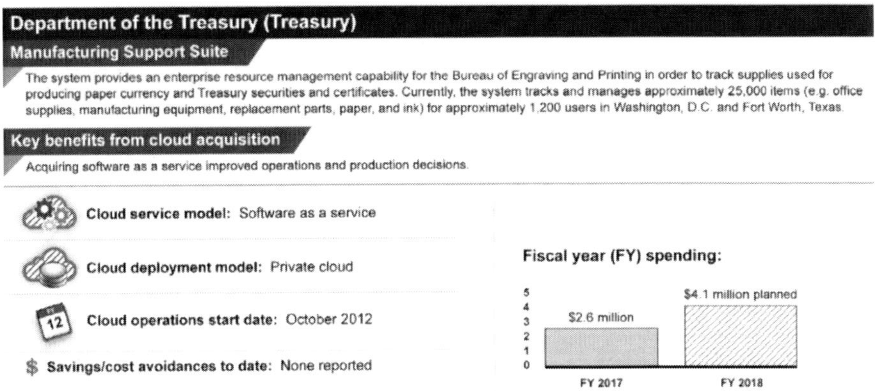

Source: GAO analysis of agency data. | GAO-19-58.

Figure 3. Treasury's Cloud Acquisition Supports Enterprise Resource Planning.

Key Benefits

Acquiring Software as a Service Improved Operations and Production Decisions

According to Bureau of Engraving and Printing officials, by consolidating the bureau's 16 legacy systems with a single system in the cloud, the bureau made significant improvements to its operations through automation, improved data quality, and increased availability, performance, and security. Specifically, acquiring software as a service enabled the bureau to purchase capabilities that its previous systems did not have. By doing so, this eliminated some manual data entry and improved the bureau's ability to, among other things, automate production decisions which allowed users to focus on more critical tasks.

For example, bureau staff reported that now they enter data from monthly Federal Reserve Bank orders into the Manufacturing Support Suite to determine the denominations and quantities of currency to produce, along with what banks will receive it. In addition, the new system uses current production times and data to determine when to replenish existing inventory and supplies, such as ink and paper, in order to prevent operational delays and downtime. Furthermore, staff stated that they now make compliance decisions using automated alerts and triggers, which help to prevent the release of products that do not meet bureau standards.

According to bureau officials, the bureau has also improved data quality and reduced the amount of time that database administrators have to spend correcting errors. For example, when users enter data into the system, real-time data validation checks prevent common errors and prompt users to make corrections before submission. In addition, staff stated that the bureau implemented bureau-specific data checks, including which accounts could be associated with item categories and cost centers, which has reduced user errors and improved the reliability of the data.

Finally, the bureau has improved system availability, security, and performance by acquiring software as a service. For example, bureau officials stated that they selected a FedRAMP-approved[73] provider and established service level agreements with the provider to help ensure system availability, security, and performance, including disaster recovery capabilities that were not available for the legacy systems. In addition, bureau staff said that they no longer need to update and maintain IT software and hardware, which has saved time and resources, and decreased system downtime.

Transportation's Cloud Acquisition Stores Public Transit Data

The Department of Transportation's (Transportation) Federal Transit Administration provides financial and technical assistance to local public transit systems, including buses, subways, light rail, commuter rail, trolleys and ferries. According to the Federal Transit Administration IT Director, in October 2014, the National Transit Database was migrated to the cloud in order to improve customer experience, mission assurance, agility and responsiveness. Previously, the legacy database had several challenges,

---

[73] FedRAMP is a government-wide program providing a standardized approach to security assessment, authorization, and continuous monitoring for cloud products and services.

including the use of obsolete technology, poor usability, and problems with data accuracy. In addition, developing new functionality for the legacy system was a lengthy process, which decreased the ability of developers to respond to other user needs.

By transitioning to the cloud, the Federal Transit Administration established a centralized access portal for users, which consolidated systems, eliminating the need to remember multiple passwords for external users, and added a single sign on feature for internal users. Staff in the Federal Transit Administration also reported that they improved the database's user interface by implementing improved system validation functionality for transit data. In addition, the cloud provided software developers with tools to develop functionality quicker to help improve the database's responsiveness to user needs. Figure 4 provides a summary of Transportation's cloud acquisition.

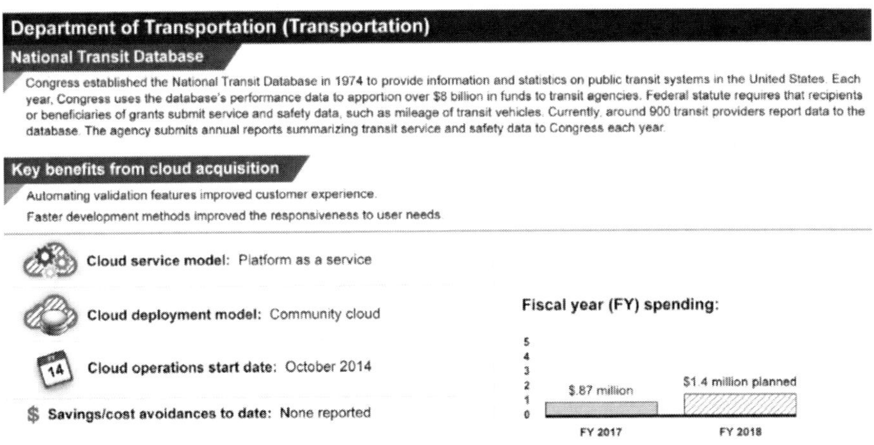

Source: GAO analysis of agency data. | GAO-19-58.

Figure 4. Transportation's Cloud Acquisition Stores Public Transit Data.

Key Benefits

Automating Validation Features Improved Customer Experience

According to Federal Transit Administration officials, by moving to the cloud, developers established automated validation features which use historical data to identify outliers and prevent potential user data entry errors. Officials reported that analysts previously performed manual data validation to ensure the accuracy of customer-entered data. The new cloud version of the National Transit Database uses historical data to identify errors and leverages cross-form data validation for the current reporting year, which has reduced the time it takes to validate the data.

Faster Development Methods Improved the Responsiveness to User Needs

According to Federal Transit Administration officials, with the deployment to the cloud, the agency adopted faster development processes, which led to more frequent releases of functionality. For example, the database's developers regularly receive requests from transit customers for enhancements that would traditionally take longer to implement in the prior legacy environment. By leveraging the cloud framework and improved Agile development procedures, officials reported that developers can now engage users earlier to make adjustments based on their feedback, thereby focusing more directly on meeting business needs.

GSA's Cloud Acquisition Enhances Federal Data Analytics

The General Services Administration (GSA) helps federal agencies build and acquire office space, products and other workspace services, and oversees the preservation of historic federal properties. According to GSA's Chief Data Officer, in 2015, GSA began developing an enterprise platform pilot program, Data to Decisions, in order to improve the agility, responsiveness, efficiency, and operations of the agency's data analytics capabilities. Previously, GSA's data analytics operations had redundancy and overlap, including similar contracts and data sources, negatively affecting data sharing efforts across the agency.

Subsequently, in October 2015, GSA's CIO, Chief Data Officer, and Chief Technology Officer moved the program to the cloud while consolidating existing contracts to create a centralized web portal. GSA officials reported that the new portal provides new data analytics capabilities for staff to use in generating analyses to advise decision makers at the agency and across other federal agencies, while also saving staff time in producing these analyses. For example, the centralized web portal allows the agency's 400 data practitioners to, among other things, build data models, understand business operations through analytics and visualizations, and publish dashboards and reporting. Figure 5 provides a summary of GSA's cloud acquisition.

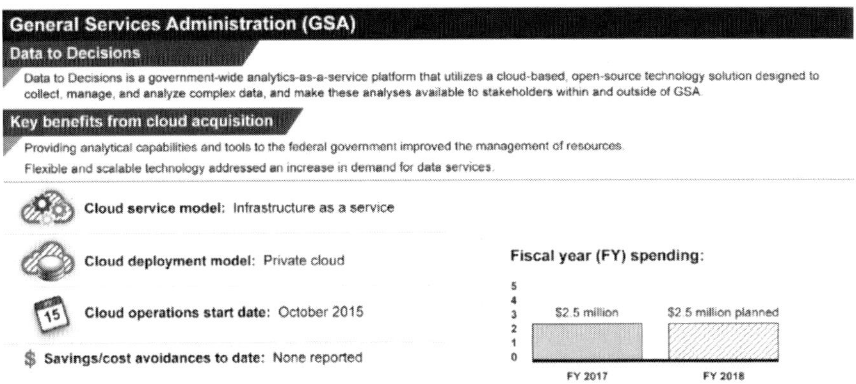

Source: GAO analysis of agency data. | GAO-19-58.

Figure 5. GSA's Cloud Acquisition Enhances Federal Data Analytics.

Key Benefits

Providing Analytical Capabilities and Tools to the Federal Government Improved the Management of Resources

According to GSA officials in the Office of the CIO, cloud deployment has enabled the sharing of data and other analytical tools across the federal government to help agencies better manage their resources and create efficiencies in data management. For example, previously, agencies did not have access to detailed data regarding agency-owned and GSA-managed property in their asset portfolios. By moving to the cloud, GSA officials reported that they developed two tools called the Real Property Management Tool and the Asset Consolidation Tool. These tools were deployed to between 30 and 40 federal agencies, which enabled these agencies to identify potential opportunities to consolidate their building properties or co-locate office spaces to help save resources. Specifically, the tools provided dashboards that showed expiring leases and occupancy agreements, as well as excess and underutilized space. Further, the program provided data to federal agencies that were not previously available. By doing so, officials said that smaller agencies did not have to invest in their own data analytics capabilities or acquire additional staff resources for data analytics.

Flexible and Scalable Technology Addressed an Increased Demand for Data Services

According to GSA officials in the Office of the CIO, as the program has expanded its data analytics capabilities, program usage has grown over time. In particular, in 2018, of the program's 7,200 users, more than 80 percent were from federal agencies other than GSA. Cloud deployment has allowed GSA to easily scale resources to manage changes in user traffic and enabled agency personnel to focus on the mission rather than managing a data center to respond to these changes in demand. For example, in 2017, GSA officials said that they sent out a notice to approximately 1 million federal employees who had completed its annual tenant satisfaction survey notifying them that the survey's results were available. As a result, several thousand users tried to access the report on the program's portal, affecting the program's operations.

Officials said that the agency was able to scale up the portal's resources and capabilities to handle the demand and then scale the resources back once user traffic returned to normal levels.

VA's Cloud Acquisition Improves Veteran Benefits and Services

The Department of Veterans Affairs (VA), among other duties, administers a variety of benefits and services that provide financial and other forms of assistance to veterans, their dependents, and survivors. According to the Deputy Assistant Secretary for Enterprise Program

Management, in March 2016, the VA CIO deployed the Vets.gov web portal to the cloud in order to improve veterans' customer experience and scale resources to meet demand. Previously, VA officials reported that they had experienced challenges with its legacy websites. Specifically, the websites were not designed using federal government web standards, including browser compatibility and accommodations for the needs of individuals with disabilities. In addition, the websites required users to remember several sets of login information to access many features on approximately 500 websites.

By moving to the cloud, VA officials stated that the agency has been able to better address veterans' needs by consolidating access to over 500 of the agency's websites for benefits and services. The new easier, mobile-friendly web portal requires only one login for all 500 websites, and incorporates features for users with disabilities, such as blind veterans.

Further, the program was able to scale up the portal's resources to meet the increased demand for online benefits and services, while adopting a design approach that better incorporated the needs of veterans and delivered functionality more quickly. In November 2018, the Vets.gov cloud platform became the building block for the agency's new homepage at VA.gov. Figure 6 provides a summary of VA's cloud acquisition.

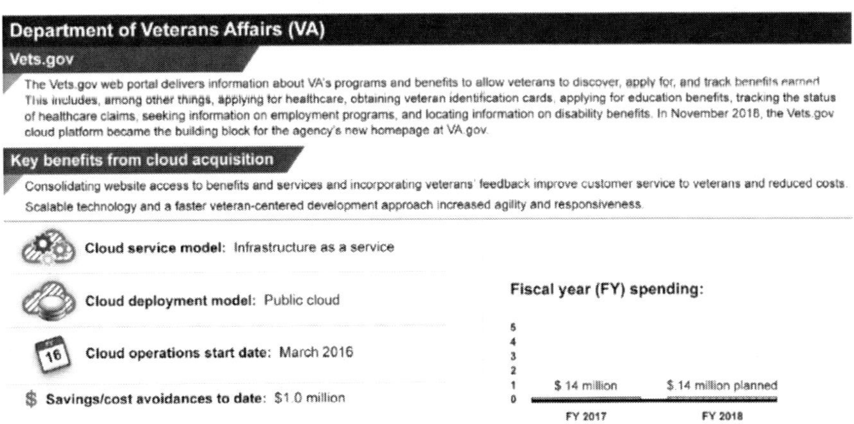

Source: GAO analysis of agency data. | GAO-19-58.

Figure 6. VA's Cloud Acquisition Improves Veteran Benefits and Services.

Key Benefits

Consolidating Website Access to Benefits and Services and Incorporating Veterans' Feedback Improved Customer Service to Veterans and Reduced Costs

According to VA officials in the Office of the CIO, by moving to the cloud, VA has worked to improve veterans' access to benefits and services through its websites in several key areas. For example, Vets.gov is intended to be mobile-friendly and work on any computing device with a compliant web browser, avoiding the need to install separate software to apply for benefits. In addition, officials stated that the agency intends the portal to be easier for veterans to search for services. For instance, VA had previously developed an application to help veterans schedule medical appointments but VA officials reported that veterans could not easily locate the application after searching across multiple VA websites.

In addition, VA officials stated that the agency was also able to reduce costs because, in moving to the cloud, Vets.gov cost the agency 85 percent less than it would have cost to build a traditionally hosted service with the same features. VA also retired a legacy application, which saved an estimated $1 million in annual contract costs.

Scalable Technology and a Faster Veteran-Centered Development Approach Increased Agility and Responsiveness

According to VA officials in the Office of the CIO, moving to the cloud allowed VA to acquire more flexible and scalable technologies in order to scale resources up and down to meet demand, while incorporating a faster, more user-friendly design approach. For example, after its launch, officials said that Vets.gov received a spike in the number of veterans that chose to submit online applications for healthcare, which the agency was able to handle by scaling up resources to meet the spike in demand.

In addition, VA officials reported that the agency adopted a design approach in the cloud that, among other things, allowed it to adopt Agile methods to more quickly deliver releases. For example, based on feedback,

VA incorporated mobile friendly design features—40 percent of Vets.gov users access benefits and services through a mobile device.

Officials said that the agency has made efforts to focus on the needs of veterans first by using an iterative design approach that incorporates user feedback into the design process so that no features in the portal are deployed without a final usability test with a veteran. VA officials also reported that using the cloud has allowed the agency to deploy new features as soon as they are ready, in small incremental daily releases.

Further, the officials noted that VA developers have worked with veterans on the portal's healthcare claims status tracker. Specifically, veterans can access the status of their healthcare claims that may be experiencing a backlog in processing, along with an estimated decision date. Lastly, officials reported that by incorporating an online application, Vets.gov reduced the number of paper-based healthcare applications submitted by veterans. In fiscal year 2018, users submitted over 750,000 digital forms for benefits through Vets.gov.

Justice's Cloud Acquisition Hosts Data Center

The Department of Justice's (Justice) U.S. Trustee Program (USTP) is responsible for overseeing the administration of bankruptcy cases and private trustees within the United States.[74] According to USTP's Chief Technology Officer, in June 2016, executives in the Program, including the CIO, decided to migrate USTP's operations to the cloud to meet regulatory requirements, reduce costs, and improve agility, efficiency, and responsiveness. Officials said that their office had conducted an evaluation and determined that, in order to fulfill OMB's mandate to consolidate agency data centers, USTP would have to spend at least $1 million for an on-premise consolidation. Officials reported that USTP also faced challenges with having adequate backup capabilities and implementing

[74] Bankruptcy cases filed in Alabama and North Carolina are overseen by a Bankruptcy Administrator from one of the judicial districts in those states rather than the U.S. Trustee Program.

new technological solutions due to its legacy computing environment and the time it took to purchase and install new hardware and software. Subsequently, in March 2017, the Program moved its operations to the cloud and avoided the cost of consolidating its data centers. In addition, officials in USTP said that the move to the cloud helped them address backup issues, and speed up the development and testing of new applications. Figure 7 provides a summary of Justice's cloud acquisition.

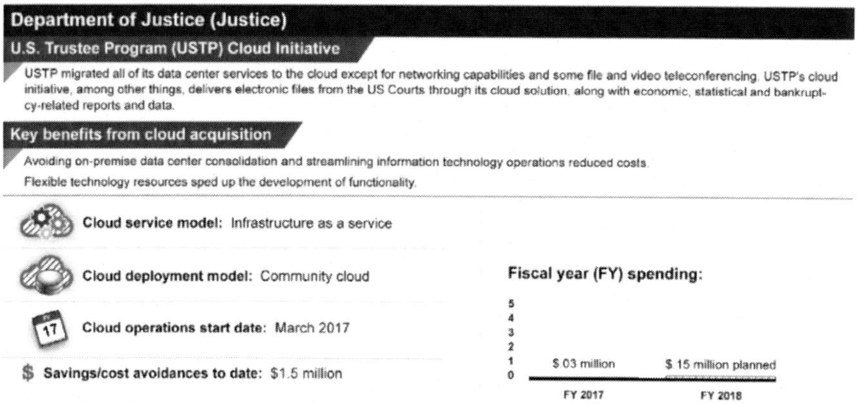

Source: GAO analysis of agency data. | GAO-19-58.

Figure 7. Justice's Cloud Acquisition Hosts Data Center.

Key Benefits

Avoiding an on-Premise Data Center Consolidation and Streamlining IT Operations Reduced Costs

According to USTP officials, by moving to the cloud, their office avoided at least $1 million in costs, while resolving internal performance issues, and streamlining the management of its contracts. Specifically, officials said that USTP shut down 1 of its 2 data centers and reduced its server inventory from 140 to 75 and the number of vendors from 21 to 9. In addition, the office eliminated an estimated 50- 70 monthly IT staff hours dedicated to resolving backup issues.

Flexible Technology Resources Sped up the Development of Functionality

According to USTP officials, acquisitions of new technology previously took several months because of the time needed to estimate requirements and wait for officials to purchase and install hardware and software. By moving to the cloud, USTP officials stated that the intention is to be able to develop and test new applications faster, and determine their viability, with minimal time and costs. Specifically, officials reported that they have set up a cloud test lab to better understand system requirements by scaling up and down resources as needed and experimenting with new capabilities. In addition, while USTP's legacy monitoring solution required consulting assistance and took months to implement, officials noted that they were able to set up a similar solution in the cloud within 1 week.

DHS's Cloud Acquisition Supports Information Sharing and Collaboration

The Department of Homeland Security (DHS) collaborates with a variety of agencies and organizations to share information related to homeland security. According to the program's Service Operations Manager, DHS's CIO migrated the Homeland Security Information Network to the cloud in July 2017 in order to improve the system's availability and operational efficiency, while reducing costs. Officials stated that, previously, the agency had faced challenges in ensuring the system's redundancy and deploying new network enhancements quickly. This was due to the costs and time frames associated with acquiring new infrastructure and maintaining and upgrading current infrastructure. In addition, the agency was not able to quickly develop and deploy new capabilities to meet user needs.

By moving to the cloud, officials stated that the agency was able to implement a disaster response capability and improve the system's operational efficiency, while also establishing more efficient environments for software development and testing. In addition, the agency was able to shut down an existing data center, which achieved cost savings of at least 30 percent from hosting the network in the data center. Figure 8 provides a summary of DHS's cloud acquisition.

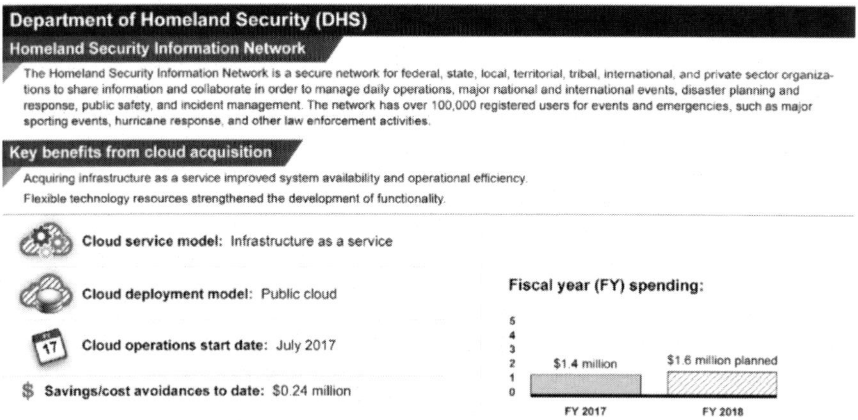

Source: GAO analysis of agency data. | GAO-19-58.

Figure 8. DHS's Cloud Acquisition Supports Information Sharing and Collaboration.

Key Benefits

Acquiring Infrastructure as a Service Improved System Availability and Operational Efficiency

According to DHS officials in the Office of the CIO, migrating to the cloud has improved the system's availability and operational efficiency, which cost less money than the prior hosting solution. For example, acquiring infrastructure as a service provided increased redundancy over the old solution and has helped to ensure the network remains continuously available for daily operations and emergency response. The officials stated that, previously, the agency had not been able to implement a disaster recovery capability because it would cost over $1.5 million to build and maintain a second active network environment. Moving to the cloud enabled the agency to implement this capability for significantly less cost.

In addition, officials in the CIO's office said that the acquisition of infrastructure as a service has enabled the agency to improve the operational efficiency of the system. For example, network managers can easily stand up new virtual hardware, networking, and storage capabilities, or make changes to existing infrastructure, in less than a day. Officials said that, previously, it used to take staff several months to make these changes manually. This allows network managers to respond very rapidly to changes in user demand, particularly if there are emergencies or natural disasters, and then scale down resources during non-use periods. For example, officials said that managers scaled up resources to support first responders from federal, state, and local governments to share weather, response and recovery information during Hurricanes Harvey, Irma, and Jose, and the West Coast wildfires. In addition, the officials noted that network managers now have access to the latest virtual hardware and the agency does not have to pay for hardware refreshments.

Flexible Technology Resources Strengthened the Development of Functionality

According to DHS officials in the Office of the CIO, moving to the cloud enabled the agency to very inexpensively build multiple environments in the system for software development, testing, and production, which has improved the development and deployment of new services. Software developers now have consistent and standardized environments, which helps to reduce the risk of errors and security vulnerabilities, as well as configuration issues. DHS officials stated that all of these issues would previously require staff time and funding to resolve. The developers can also now use automation tools to deploy new code from development into production more quickly to help meet user needs for new functionality. In addition, officials noted that cloud providers are constantly adding new services that users can leverage to do their work more efficiently, without the time and cost of the agency having to develop or procure this capability separately.

Agriculture's Cloud Acquisition Improves Enterprise Content and Electronic Records Management

The Department of Agriculture's (Agriculture) U.S. Forest Service manages 193 million acres of federal land in order to sustain the health, diversity, and productivity of the nation's forests and grasslands for present and future generations. According to the Acting Assistant Forest Service CIO for Natural Resources and Environment, in August 2017, the Forest Service began deploying a new enterprise content management and electronic records management system, called Pinyon, to the cloud to help improve operations and the management of electronic records. The move also addressed federal requirements related to electronic records management.[75] Officials stated that, previously, the Forest Service relied on a shared storage drive for enterprise content management. Officials reported that this drive was highly proprietary, slow, unreliable, and a security vulnerability because it could not be easily maintained. In addition, officials reported that the shared storage drive was on the verge of failure because the vendor no longer supported and upgraded the system.

By acquiring two software as a service solutions for enterprise content and electronic records management, officials said that the Forest Service was able to quickly deploy a new system with only some limited software customization for the integration of the two solutions. The Forest Service completed this in two phases; officials deployed the enterprise content management solution in August 2017 and the electronic records management solution began deployment in August 2018. Officials reported that they plan to fully deploy the system by December 2018. Figure 9 provides a summary of Agriculture's cloud acquisition.

---

[75] By December 31, 2019, OMB and the National Archives and Records Administration will require federal agencies to manage all permanent electronic records to the fullest extent possible in an electronic format in order to transfer those records to the archives. See OMB and National Archives and Records Administration, *Memorandum for the Heads of Executive Departments and Agencies and Independent Agencies: Managing Government Records Directive*, M-12-18 (Washington, D.C.: Aug. 24, 2012).

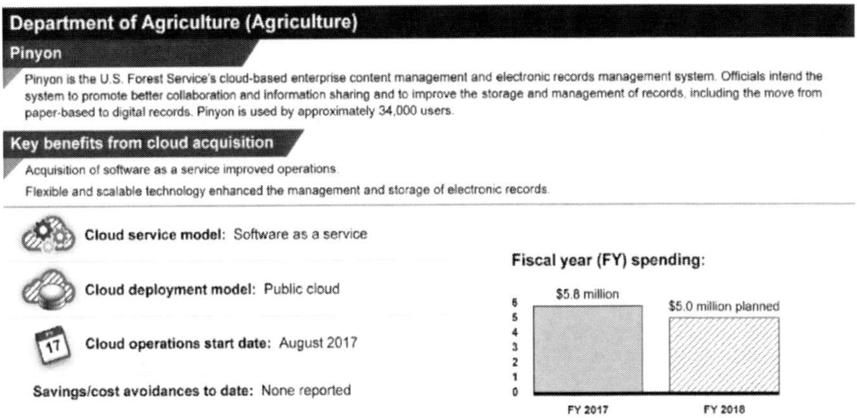

Source: GAO analysis of agency data. | GAO-19-58.

Figure 9. Agriculture's Cloud Acquisition Improves Enterprise Content and Electronic Records Management.

Key Benefits

Acquisition of Software as a Service Improved Operations

According to Forest Service officials in the Office of the CIO, by acquiring software as a service, the Forest Service was able to implement new enterprise content management capabilities and collaboration tools quickly without the costs and risks associated with software development. Officials said that, previously, users did not have capabilities for managing their own content such as setting permissions, granting access privileges to documents, or easily managing different document versions. In addition, officials noted that users relied heavily on email to collaborate on daily work activities as other collaboration tools were not available. By acquiring software as a service, officials said that the Forest Service was able to quickly implement enhanced workflow and document management capabilities and add new tools for collaboration, which has increased staff productivity.

Furthermore, acquiring software as a service allowed the Forest Service to integrate their new system with Agriculture's electronic authentication system, which the agency could not previously accomplish with the legacy system. By integrating these systems, Forest Service officials said that the agency has increased the accessibility of the Forest Service's information by allowing staff to securely access files regardless of physical location.

Going forward, officials in the Forest Service said that they are exploring other features and capabilities offered by the cloud vendor to help better meet mission needs. For example, the Forest Service regularly collaborates with a variety of other agencies, state and local governments, educational institutions and other organizations on issues related to managing federal lands and responding to natural disasters, such as wildfires. Officials noted that the Forest Service hopes to use shared virtual workspaces and other collaboration tools to engage these partners.

In addition, by acquiring software as a service, Forest Service officials said that they have ensured that there is a system in place that the vendor will automatically upgrade with new enhancements, capabilities, and the latest technology. For example, in order to meet new federal cybersecurity requirements, Forest Service officials said that they have been able to work with the cloud vendor to ensure the vendor incorporates software changes to meet these requirements.

Flexible and Scalable Technology Enhanced the Management and Storage of Electronic Records

According to Forest Service officials in the Office of the CIO, by moving to the cloud, the Forest Service was able to acquire new storage capabilities that are easily scalable as its volume of electronic records grows over time. Officials said that, previously, the Forest Service used both paper-based records and a shared storage drive for storing work documents and other operational records.

Paper- based records were stored in file cabinets and warehouses while the shared storage drive maintained approximately 320 million files and 250 terabytes of data. In addition, the agency previously used tape backups for the shared storage drive. By moving to the cloud, officials in the Forest Service said that they gained unlimited storage and electronic backup capabilities. Further, Forest Service officials said the new system is intended to be able to easily scale up storage resources as needed for the digitization of its paper-based records and handle the future volumes of electronic records.

In addition, by acquiring software as a service, officials in the CIO's office reported that the Forest Service was able to meet the federal requirement for electronic records management more than a year before the December 2019 deadline, which the prior shared drive could not meet.

Commerce's Cloud Acquisition Enhances Access to Weather Data

The Department of Commerce's (Commerce) National Oceanic and Atmospheric Administration works to understand and predict changes in climate, weather, oceans, and coasts, and shares that knowledge and information with others. According to Commerce's Acting Chief Information Officer, in September 2017, the National Oceanic and Atmospheric Administration's CIO and National Weather Service leadership decided to deploy its public weather websites to the cloud in order to improve the agility and responsiveness of these websites in a cost-effective manner. Officials stated that, previously, in 2016, as a result of Hurricane Matthew, hundreds of millions of web requests led to failures with the program's on-premise infrastructure, causing websites to become unavailable to the public for a period of time. Subsequently, in September 2017, prior to the landfall of Hurricane Irma, officials stated that the agency launched its weather cloud content delivery network. This new network is intended to ensure the availability of weather-related information, while avoiding the additional expenses for infrastructure that would likely go unused during normal business operations. Figure 10 provides a summary of Commerce's cloud acquisition.

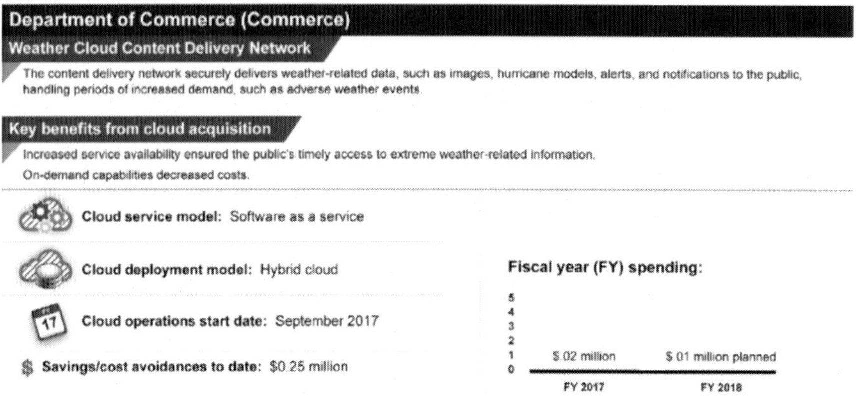

Source: GAO analysis of agency data. | GAO-19-58.

Figure 10. Commerce's Cloud Acquisition Enhances Access to Weather Data.

Key Benefits

Increased Service Availability Ensured the Public's Timely Access to Extreme Weather-Related Information

According to National Oceanic and Atmospheric Administration officials, the deployment of the weather cloud content delivery network in September 2017 helped websites handle the web requests for data on Hurricanes Irma and Maria by scaling up the resources needed to handle the increased requests. Normally, the weather websites receive approximately 26 million daily web requests from the public. However, officials noted that the number of requests increases dramatically during adverse weather events, such as hurricanes. For example, officials said that in August 2017, the websites began experiencing delays because of the high volume of hurricane- related requests from Hurricane Harvey—including approximately 218 million web requests on August 31, 2017 alone. After deployment to the cloud in September 2017, officials reported that over the course of two days, the weather cloud content delivery network successfully scaled up its resources and handled approximately two billion web requests received through the administration websites.

On-Demand Capabilities Decreased Costs

According to National Oceanic and Atmospheric Administration officials, by acquiring software as a service, it avoided the cost of expanding existing on-premise infrastructure to handle sudden surges in demand that only last a short period of time, as well as associated maintenance costs. Officials said that the program can now scale up the resources supporting the weather cloud content delivery network whenever it anticipates an adverse weather event that would lead to greater demand for website information.

Defense's Cloud Acquisition Enhances Transportation Command Systems

The Department of Defense's (Defense) U.S. Transportation Command (USTRANSCOM) provides common user and commercial air, land, and sea transportation, as well as terminal management and air refueling, in support of the military's deployment, employment, sustainment, and re- deployment efforts. USTRANSCOM's Chief of Cyber Operations and Readiness Division reported that in January 2017, the USTRANSCOM Commander made the decision to migrate all of the command's systems to the cloud in order to improve mission assurance, agility, responsiveness, efficiency, and operations. Officials reported that, previously, the command had experienced a massive power outage affecting the availability of approximately 25 legacy systems that lacked the capability to quickly recover from network failures. In addition, officials noted that the command's system, used to manage world-wide moves of Defense personnel property, was not user-friendly, and was difficult to maintain because the agency built the system using waterfall software development methods. Lastly, officials said that the command had largely relied on manual reporting activities that took numerous staff hours to produce to make financial, operational, planning, and support decisions.

By beginning to transition to the cloud in January 2018, USTRANSCOM officials said that the command is in the process of ensuring its systems are secure and continuously available, and is developing capabilities to improve the usability of its legacy systems. In addition, officials reported that the command is streamlining its tracking and reporting mechanisms to allow users to automatically generate key reports, which will give decision makers access to more current and accurate information to help improve program operations. USTRANSCOM officials said that executive sponsorship is absolutely critical for migrating to the cloud to overcome culture change by bringing together people throughout the enterprise. In addition, the command's cloud center of excellence team facilitates the command's adoption of cloud by, among other things, training users and addressing governance issues. Figure 11 provides a summary of Defense's cloud acquisition.

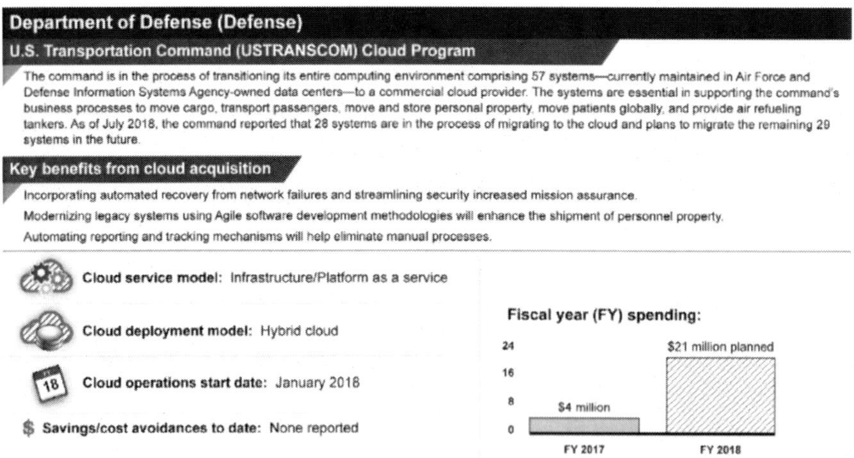

Source: GAO analysis of agency data. | GAO-19-58.

Figure 11. Defense's Cloud Acquisition Enhances Transportation Command Systems.

Key Benefits

Incorporating Automated Recovery from Network Failures and Streamlining Security Increased Mission Assurance

According to USTRANSCOM officials, by moving the command's network to the cloud, the command has been able to design and build its new network with higher levels of availability. For example, officials said that if a network segment becomes unavailable, the cloud technology has the capability to automatically reroute traffic to help reduce the amount of delay that users experience. In addition, officials reported that developers have been working to automate several hundred security checks that are part of Defense's security technical implementation guides[76] by implementing a repeatable, automated process instead of doing manual checks. Officials noted that, previously, manually checking of the status of configurations would take hundreds of man hours to complete. Eventually, the command anticipates that automation will save these hours of manual checks. The command plans to implement the new capability in May 2019.

Replacing a Legacy System with a Cloud-Based System Developed Using Agile Software Development Methodologies Will Enhance the Shipment of Personnel Property

According to USTRANSCOM officials, moving to the cloud has assisted the command by replacing the legacy system that manages moves of Defense personnel property, like household goods, with a mobile prototype built in the cloud. Officials reported that the legacy system currently uses a variety of commercial products that are difficult to maintain and do not efficiently address the command's complex business processes for personnel property moves, all of which affects the usability of the system. Currently, the command is using Agile software development methodologies to reengineer its business processes to develop

[76] Defense security technical implementation guides, developed by the Defense Information Systems Agency, provide the required security configuration settings for Defense devices and systems related to information assurance in order to securely configure these devices and systems. In addition, each guide also has a checklist that is used to review and certify that the appropriate configurations are in place.

a solution that is mobile and user-friendly. The new mobile prototype is intended to allow personnel to request access in order to manage the moves of certain household goods.

Officials reported that the command initially planned to deploy the prototype in June 2018 but deployment was delayed and a new date had not yet been identified.

Automating Reporting and Tracking Mechanisms Will Help Eliminate Manual Processes

According to USTRANSCOM officials, the command is in the process of automating its processes for reporting and tracking cargo shipments utilizing cloud technologies. Currently, the command employs manual processes to track and monitor a variety of its cargo shipments. For example, officials reported that five analysts typically spend one day compiling a status report that details delays with food shipments for Defense military exercises and operations. In addition, analysts currently have to query up to 11 Defense and commercial carrier systems to compile a report on high-priority shipments across the combatant commands. Officials noted that these analysts often experience delays getting access to timely information and must also resolve conflicting information in various transportation systems. However, with the transition to cloud services, officials in USTRANSCOM reported that analysts will have the capability to automatically generate reports based on defined criteria, such as shipment method or destination, and use data feeds that officials can continuously update. By developing phase one of the system in the cloud in fiscal year 2018, officials reported that they will be able to monitor delays in a shipment and immediately take action to change the mode of transportation or source shipments from alternate suppliers. The command plans to release the full operational capability in fiscal year 2020, which, officials noted, will give authorized users near real-time access to shipment information, including estimates of whether a shipment will arrive on time.

Appendix VII: Comments from the Department of Commerce

UNITED STATES DEPARTMENT OF COMMERCE
The Secretary of Commerce
Washington, D.C. 20230

February 13, 2019

Ms. Carol C. Harris
Director, Information Technology Acquisition
Management Issues
U.S. Government Accountability Office
441 G Street, NW
Washington, DC 20548

Dear Ms. Harris:

Thank you for the opportunity to review and comment on the Government Accountability Office's (GAO) draft report titled "*Cloud Computing: Agencies Have Increased Usage and Realized Benefits, but Cost and Savings Data Need to be Better Tracked* (GAO-19-58)."

On behalf of the Department of Commerce, I have enclosed our comments on the draft report. We concur with the recommendation and will take steps to implement it.

Sincerely,

Wilbur Ross

Department of Commerce's Comments on

GAO Draft Report titled *Cloud Computing: Agencies Have Increased Usage and Realized Benefits, but Cost and Savings Data Need to be Better Tracked*

(GAO-19-58)

The Department of Commerce has reviewed the draft report and offers the following comments for consideration.

General Comments

The report on documenting cost and savings data thoroughly assesses the challenges of collecting and accurately reporting savings derived from the implementation of Cloud Computing. The report's discussion of the various challenges with identifying savings, reporting details, and identifying investment types that do not provide savings is generally well-informed, thorough, and balanced.

Comments on Recommendation

The Government Accountability Office (GAO) made one (1) recommendation to the Department of Commerce in the report.

- **Recommendation:** The Secretary of Commerce should ensure that the CIO of Commerce establishes a consistent and repeatable mechanism to track savings and cost avoidances from the migration and deployment of cloud services.

  Commerce Response: The Department of Commerce agrees with this recommendation.

  Commerce will develop and implement a consistent and repeatable process in accordance with the recommendation in the report and will apply this to all Cloud Computing investments.

APPENDIX VIII: COMMENTS FROM THE DEPARTMENT OF EDUCATION

UNITED STATES DEPARTMENT OF EDUCATION

OFFICE OF THE CHIEF INFORMATION OFFICER

THE CHIEF INFORMATION OFFICER

January 15, 2019

Ms. Carol C. Harris
Director, Information Technology
Acquisition Management Issues
Government Accountability Office
441 G Street, NW
Washington, DC 20548

Dear Ms. Harris:

I am pleased to provide the U.S. Department of Education's (ED's or Department's) response to the Government Accountability Office's (GAO's) draft report GAO-19-58, *Cloud Computing: Agencies Have Increased Usage and Realized Benefits, but Cost and Savings Data Need to be Better Tracked.* We understand that GAO conducted this audit to review the Department's use of cloud computing services and the extent to which the Department tracked cost savings or avoidance data when it used such services. We appreciate the opportunity to respond to the three recommendations for ED in the GAO draft report.

GAO Recommendation 7: The Secretary of Education should ensure that the CIO of Education establishes guidance on assessing new and existing IT investments for suitability for cloud computing services, in accordance with OMB guidance.

Response: The Department concurs with the GAO recommendation. The Department is currently developing guidance on assessing new and existing IT investments for suitability for cloud computing services.

GAO Recommendation 8: The Secretary of Education should ensure that the CIO of Education completes an assessment of all IT investments for suitability for migration to a cloud computing service, in accordance with OMB guidance.

Response: The Department concurs with the GAO recommendation. The Department will develop guidance and ensure that the CIO of Education completes an assessment of all IT investments for suitability for migration to a cloud computing service.

GAO Recommendation 9: The Secretary of Education should ensure that the CIO of Education establishes a consistent and repeatable mechanism to track savings and cost avoidances from the migration and deployment of cloud services.

Response: The Department concurs with the GAO recommendation. The Department is currently developing guidance that will establish a consistent and repeatable mechanism to track savings and cost avoidances from the migration and deployment of cloud services.

Sincerely,

Jason K. Gray

APPENDIX IX: COMMENTS FROM THE DEPARTMENT OF ENERGY

Department of Energy
Washington, DC 20585

1/25/2019

Ms. Carol C. Harris
Director, Information Technology and Management Issues
U.S. Government Accountability Office
441 G Street, N.W.
Washington, D.C. 20548

Dear Ms. Harris:

Thank you for the opportunity to provide the Department of Energy's (DOE or Department) management response to the Government Accountability Office's (GAO) draft report *Cloud Computing: Agencies Have Increased Usage and Realized Benefits, but Cost and Savings Data Need to be Better Tracked* (GAO-19-58). GAO conducted this audit to (1) evaluate DOE's progress in implementing cloud services and determine the extent to which DOE has experienced cost savings and (2) describe reported examples of successful cloud migrations and deployments.

DOE concurs with GAO's report recommendations and provides details concerning the Department's responses in the Enclosure.

Sincerely,

Stephen (Max) Everett
Chief Information Officer

MANAGEMENT RESPONSE
GAO Draft Report, GAO-19-58
CLOUD COMPUTING:
Agencies Have Increased Usage and Realized Benefits, but Cost and Savings Data Need to be Better Tracked (Job Code 102299)

Recommendation 10: The Secretary of Energy should ensure that the CIO of Energy updates the agency's guidance on assessing IT investments for suitability for cloud computing services to include a requirement to assess new acquisitions for these services.

Management Decision: Concur

DOE is establishing an enterprise cloud migration approach to assist with cloud services adoption. Consistent with the Cloud First policy, DOE elements are required to use cloud infrastructure wherever possible when planning new mission or support applications or consolidating existing applications. Updates to the agency's IT budget guidance will address assessing IT investments for suitability for cloud computing services.

Estimated Completion Date: December 30, 2019

Recommendation 11: The Secretary of Energy should ensure that the CIO of Energy completes an assessment of all IT investments for suitability for migration to a cloud computing service, in accordance with OMB guidance.

Management Decision: Concur

DOE is establishing an enterprise cloud migration approach to assist with cloud services adoption. Consistent with the Cloud First policy, DOE elements are required to use cloud infrastructure wherever possible when planning new mission or support applications or consolidating existing applications. In compliance with FITARA and OMB guidance, the agency's IT budget guidance will address assessing IT investments for suitability for cloud computing services.

Estimated Completion Date: December 30, 2019

Recommendation 12: The Secretary of Energy should ensure that the CIO of Energy establishes a consistent and repeatable mechanism to track savings and cost avoidances from the migration and deployment of cloud services.

Management Decision: Concur

The CIO has established a consistent and repeatable mechanism to track savings and cost avoidances. The CIO will continue to refine the process to improve tracking savings and cost avoidances from the migration and deployment of cloud services.

Estimated Completion Date: December 30, 2019

Appendix X: Comments from the Health and Human Services

DEPARTMENT OF HEALTH & HUMAN SERVICES

OFFICE OF THE SECRETARY

Assistant Secretary for Legislation
Washington, DC 20201

JAN 17 2019

Carol C. Harris
Director, Information Technology Acquisition
Management Issues
U.S. Government Accountability Office
441 G Street NW
Washington, DC 20548

Dear Ms. Harris:

Attached are comments on the U.S. Government Accountability Office's (GAO) report entitled, "*Cloud Computing: Agencies Have Increased Usage and Realized Benefits, but Cost and Savings Data Need to be Better Tracked*" (GAO-19-58).

The Department appreciates the opportunity to review this report prior to publication.

Sincerely,

Matthew D. Bassett
Assistant Secretary for Legislation

GENERAL COMMENTS FROM THE DEPARTMENT OF HEALTH & HUMAN SERVICES ON THE GOVERNMENT ACCOUNTABILITY OFFICE'S DRAFT REPORT ENTITLED - CLOUD COMPUTING: AGENCIES HAVE INCREASED USAGE AND REALIZED BENEFITS, BUT COST AND SAVINGS DATA NEED TO BE BETTER TRACKED (GAO-19-58)

The U.S. Department of Health & Human Services (HHS) appreciates the opportunity from the Government Accountability Office (GAO) to review and comment on this draft report.

Recommendation 13
The Secretary of HHS should ensure that the chief information officer (CIO) of HHS establishes guidance on assessing new and existing information technology (IT) investments for suitability for cloud computing services, in accordance with Office of Management and Budget (OMB) guidance.

HHS Response
HHS concurs with GAO's recommendation.

HHS published the "HHS Cloud Computing Strategy" and will revise this document to reflect current OMB guidance and current technology environment. HHS Office of the Chief Information Officer plans to revise and update this document by the close of FY2019.

Recommendation 14
The Secretary of HHS should ensure that the CIO of HHS completes and assessment of all IT investments for suitability for migration to cloud computing services, in accordance with OMB guidance.

HHS Response
HHS concurs with GAO's recommendation.

HHS CIO will ensure all IT investments complete a review of suitability for migration to cloud computing services as part of the annual Operating Division IT portfolio reviews for FY2021. All reviews of suitability for migration to cloud computing services will be done in conjunction with IT investment business owners.

Recommendation 15
The Secretary of HHS should ensure that the CIO of HHS establishes a consistent and repeatable mechanism to track savings and cost avoidance from the migration and deployment of cloud services.

HHS Response
HHS concurs with GAO's recommendation.

HHS currently tracks all cost savings and avoidance initiatives through the OMB Integrated Data Collection. This includes savings and cost avoidance from cloud services deployments and migration. As part of the annual Operating Division IT portfolio reviews, the HHS CIO will ensure that all IT investments with cloud components are tracking any attributable costs savings or avoidance.

APPENDIX XI: COMMENTS FROM THE DEPARTMENT OF HOMELAND SECURITY

U.S. Department of Homeland Security
Washington, DC 20528

Homeland Security

March 5, 2019

Carol C. Harris
Director, Information Technology
Acquisition Management Issues
U.S. Government Accountability Office
441 G Street, NW
Washington, DC 20548

Re: Management Response to Draft Report GAO-19-58, "CLOUD COMPUTING: Agencies Have Increased Usage and Realized Benefits, but Cost and Savings Data Need to be Better Tracked"

Dear Ms. Harris:

Thank you for the opportunity to review and comment on this draft report. The U.S. Department of Homeland Security (DHS) appreciates the U.S. Government Accountability Office's (GAO) work in planning and conducting its review and issuing this report.

The Department is pleased to note GAO's positive recognition that implementing Cloud computing services has saved the Federal government hundreds of millions of dollars and realized notable benefits. For DHS, this has included improving the Department's ability to quickly respond to mission needs by providing teleworking capabilities for staff. DHS remains committed to leveraging Cloud computing technologies, migrating appropriate systems to the Cloud, and optimizing our remaining data centers[1].

In 2018, DHS established the Cloud Steering Group (CSG), which is chaired by the Under Secretary for Management and has senior level participation from DHS Components. The CSG oversees implementation of a federated, enterprise-wide strategy for accelerating the modernization and migration of our IT applications and infrastructure to the Cloud and optimizing our remaining data centers by aligning their capabilities and economics, to the extent possible, with the Cloud. DHS is implementing Cloud services in a hybrid IT, multi-cloud, federated and vendor-neutral approach.

---

[1] DHS adopted the National Institute of Standards and Technology (NIST) definitions of the Cloud, including Public, Community, Hybrid, and Private Clouds. DHS expects to support all three Cloud Computing models – Infrastructure, Platform, and Software as a Service (IaaS, PaaS, and SaaS).

DHS's robust, multi-year Component-led Cloud implementation endeavor is continually progressing. As of December 2018, a total of 171 of 628 (26 percent) systems are either planned for, in development for, actively migrating to, or operational in the Cloud, with expectations of further growth in fiscal years 2019 and 2020.

The draft report contained 2 recommendations with which the Department concurs. Attached find our detailed response to the recommendations. Technical comments were previously provided under separate cover.

Again, thank you for the opportunity to review and comment on this draft report. Please feel free to contact me if you have any questions. We look forward to working with you again in the future.

Sincerely,

JIM H. CRUMPACKER, CIA, CFE
Director
Departmental GAO-OIG Liaison Office

Attachment: DHS Management Response to Recommendations Contained in GAO-19-58

GAO recommended that the Secretary of Homeland Security ensure that the DHS Chief Information Officer:

Recommendation 16: Completes an assessment of all IT investments for suitability for migration to a Cloud computing service, in accordance with OMB guidance.

Response: Concur. DHS is continually completing comprehensive assessments of IT investments. DHS's Office of the Chief Information Officer (OCIO) Business Management Office (BMO) 2018 annual infrastructure study process compiled data regarding DHS-hosted systems and Cloud usage and migration. At this time, DHS OCIO is working to complete an analysis on the data and produce a baseline report which will include suitability for migration to a Cloud service provider. Estimated Completion Date (ECD): July 31, 2019.

Recommendation 17: Establishes a consistent and repeatable mechanism to track savings and cost avoidances from the migration and deployment of Cloud services.

Response: Concur. To date, DHS efforts to track Cloud services has been associated with planned spending at a program level. Cost savings and cost avoidance for commodities is currently collected at an aggregate level by Components for quarterly submission to OMB, but more granularity is needed for better accuracy and consistency across Components.

As DHS implements Technology Business Management for actual IT expenditures over the next several years, more granular data will become available for increased visibility of actual Cloud spending. In the meantime, DHS BMO will evaluate its current formula and processes to collect and assess actual Cloud spending by the end of FY 2019. This will support informed decisions to calculate and report cost savings or cost avoidance at a program or system level in FY 2020. Longer term, DHS will conform to any updated OMB guidance regarding data standards, reporting processes, and quality assurance mechanisms for Cloud cost savings or cost avoidance to ensure consistency across Federal agencies. ECD: September 30, 2020.

APPENDIX XII: COMMENTS FROM THE DEPARTMENT OF STATE

United States Department of State
Comptroller
Washington, DC 20520

FEB 1 2 2019

Thomas Melito
Managing Director
International Affairs and Trade
Government Accountability Office
441 G Street, N.W.
Washington, D.C. 20548-0001

Dear Mr. Melito:

We appreciate the opportunity to review your draft report, "CLOUD COMPUTING: Agencies Have Increased Usage and Realized Benefits, but Cost and Savings Data Need to be Better Tracked" GAO Job Code 102299.

The enclosed Department of State comments are provided for incorporation with this letter as an appendix to the final report.

Sincerely,

Jeffrey C. Mounts (Acting)

Enclosure:
As stated

cc: GAO – Carol C. Harris
IRM – Karen Mummaw (Acting)
OIG - Norman Brown

Department of State Response to GAO Draft Report

CLOUD COMPUTING: Agencies have increased usage and Realized Benefits, but Cost Savings Data Need to be Better Tracked (GAO-19-58, GAO Code 102299)

The Department of State appreciates the opportunity to comment on GAO's draft *"Cloud Computing: Agencies Have Increased Usage and Realized Benefits, but Cost Savings Data Need to be Better Tracked."*

The Secretary of State should ensure that the CIO of State establishes a consistent and repeatable mechanism to track savings and cost avoidances from the migration and deployment of cloud service **[Recommendation 23]**

Response:

The Department of State acknowledges and concurs with the Government Accountability Office's (GAO) recommendation that the Department needs to establish a system to track savings and cost avoidance resulting in the deployment of cloud services. To that end Business Management and Planning (BMP) was approved to hire several individuals to form a cost estimation group within BMP's Strategic Planning Office in Fiscal Year (FY) 19. After these individuals are brought on board they will be tasked to develop the project concept to create a tracking system that will follow the Technology Business Management (TBM) framework that is currently in development. BMP hopes to have a prototype system ready for testing at the end of FY19/beginning of FY20.

Appendix XIII: Comments from the Department of Transportation

U.S. Department of Transportation
Office of the Secretary of Transportation

1200 New Jersey Avenue, SE
Washington, DC 20590

FEB 13 2019

Carol C. Harris
Director, Information Technology Acquisition Management Issues
U.S. Government Accountability Office (GAO)
441 G Street NW
Washington, DC 20548

Dear Ms. Harris:

The Department of Transportation (DOT) is committed to improving its management of Information Technology (IT) services through cloud computing. The Department is leveraging its DestinationsDIGITAL initiative— an initiative to modernize DOT's technology infrastructure—to further refine and improve IT cloud computing practices within the enterprise. In addition, DOT is centralizing the procurement of non-FAA IT cloud computing services by leveraging the Department's Enterprise Cloud Services (ECS) initiative, a component of the DestinationsDIGITAL initiative.

Examples of the Department's completed and ongoing ECS actions include the following:

- established guidelines, standards and contracts for managing and overseeing DOT's cloud environments throughout the Department;
- enhanced non-FAA IT Cloud Computing inventory, allowing for greater knowledge and oversight of IT cloud activities transpiring throughout the enterprise; and
- developing a DOT-managed cloud environment where modal applications are hosted and new applications are developed. We are designing the environment to ensure each tenant application inherits department-defined, as well as the Federal Risk and Authorization Management Program-approved security controls, monitoring capabilities, development toolsets and other support services.

In addition, the Department has diligently worked to identify efficiencies generated by IT cloud computing services, and reports all such efficiencies in accordance with existing OMB guidance. The Department will work to improve reporting consistent with any updates to OMB guidance as recommended by GAO in Recommendation 1 of the draft report.

Sincerely,

Keith Washington
Deputy Assistant Secretary for Administration

APPENDIX XIV: COMMENTS FROM THE DEPARTMENT OF VETERANS AFFAIRS

THE SECRETARY OF VETERANS AFFAIRS
WASHINGTON

January 17, 2019

Ms. Carol C. Harris
Director
Information Technology
Acquisition Management Issues
U.S. Government Accountability Office
441 G Street, NW
Washington, DC 20548

Dear Ms. Harris:

The Department of Veterans Affairs (VA) has reviewed the Government Accountability Office (GAO) draft report: ***"CLOUD COMPUTING: Agencies Have Increased Usage and Realized Benefits, but Cost and Savings Data Need to be Better Tracked"*** (GAO-19-58). The enclosure sets forth the actions to be taken to address the draft report recommendations.

VA appreciates the opportunity to comment on your draft report.

Sincerely,

Robert L. Wilkie

Department of Veterans Affairs (VA) Comments to
Government Accountability Office (GAO) Draft Report
"CLOUD COMPUTING: Agencies Have Increased Usage and Realized Benefits, but Cost and Savings Data Need to be Better Tracked"
(GAO-19-58)

Recommendation 1: The Secretary of Veterans Affairs should ensure that the CIO of VA completes an assessment of all IT investments for suitability for migration to a cloud computing service, in accordance with OMB guidance.

VA Comment: Concur. The Chief Information Officer has directed that all current and new applications migrate to or be developed in the VA Enterprise Cloud. VA's Office of Information and Technology (OIT) Enterprise Cloud Solutions Office (ECSO) is working to assess applications in accordance with leadership priorities, and with incoming migration requests from projects. The assessment process, started in Fiscal Year (FY) 2018, is underway. VA is conducting the migration to the cloud in an agile fashion through quarterly wave builds, with an assessment phase built into each wave. Typically, VA assesses at least 20 applications per quarter, with at least 10 applications, which are then scheduled into the wave planning for cloud migration. VA is committed to completing the migration of 350 cloud-suitable applications to the VA Enterprise Cloud by September 30, 2024. It is anticipated that the remaining applications will be replaced/decommissioned (as reported to the Office of Management and Budget). The final quarterly assessment for the final wave is targeted to end by June 30, 2024.

Recommendation 2: **The Secretary of Veterans Affairs should ensure that the CIO of VA establishes a consistent and repeatable mechanism to track savings and cost avoidances from the migration and deployment of cloud services.**

VA Comment: Concur. ECSO and OIT's Information Technology (IT) Resource Management Office are populating the Apptio IT financial management application. Apptio IT is the tool OIT uses to track overall IT spending and cost savings associated with migration to the cloud from traditional data centers, and for instituting a scorecard to track the cloud versus traditional hosting costs. The scorecard accumulates savings and cost avoidances, including the Data Center Consolidation effort, and will capture major savings and cost avoidances once VA usage of external data centers concludes and VA data centers are consolidated. By September 30, 2020, VA expects sufficient data center consolidation/closure savings (to date, leadership has directed ECSO support of migration of VA applications of three data centers, two of which are external, by the end of FY 2020).

APPENDIX XV: COMMENTS FROM THE GENERAL SERVICES ADMINISTRATION

GSA

The Administrator

January 28, 2019

The Honorable Gene L. Dodaro
Comptroller General of the United States
U.S. Government Accountability Office
Washington, DC 20548

Dear Mr. Dodaro:

The U.S. General Services Administration (GSA) appreciates the opportunity to review and comment on the Government Accountability Office's draft report entitled *Cloud Computing: Agencies Have Increased Usage and Realized Benefits, but Cost and Savings Data Need to be Better Tracked* (GAO-19-58).

There is one recommendation addressed to the Administrator of General Services:

- The Administrator of General Services should ensure that the Chief Information Officer of GSA establishes a consistent and repeatable mechanism to track savings and cost avoidances from the migration and deployment of cloud services (Recommendation 31).

Sincerely,

Emily W. Murphy

Emily W. Murphy
Administrator

cc: Ms. Carol C. Harris, Director, Information Technology Acquisition Management Issues, GAO

APPENDIX XVI: COMMENTS FROM THE DEPARTMENT OF LABOR

U.S. Department of Labor

Office of the Assistant Secretary
for Administration and Management
Washington, D.C. 20210

JAN 2 3 2019

Ms. Carol C. Harris
Director, Information Technology
Acquisition Management Issues
Government Accountability Office
441 G Street, NW
Washington, D.C. 20548

Dear Ms. Harris:

Thank you for the opportunity to review and comment on draft report GAO-19-58 *Cloud Computing: Agencies Have Increased Usage and Realized Benefits, but Cost and Savings Data Need to be Better Tracked.* We appreciate the Government Accountability Office's (GAO) efforts and insights.

Recommendation 20: *The Secretary of Labor should ensure that the CIO of Labor updates the agency's guidance on assessing IT investments for suitability for cloud computing services to include a requirement to assess existing investments for these services.*

DOL Response: DOL concurs with the GAO recommendation and has taken steps to ensure DOL agencies include an assessment of cloud computing suitability as they move forward with their investments. This process has been integrated into our budgeting process, and agencies must submit their spend plan, including cloud feasibility assessments, annually.

Recommendation 21: *The Secretary of Labor should ensure that the CIO of Labor completes an assessment of all IT investments for suitability for migration to a cloud computing service, in accordance with OMB guidance.*

DOL Response: DOL concurs with the GAO recommendation. DOL has undertaken an initiative to review the full suite of data center based applications for cloud suitability and reinvestment. This comprehensive review is underway, and will be accomplished using a two-pronged strategy. The first of these will analyze the most problematic and out-of-date line of business applications in the inventory. The second approach will focus on the applications that are most ready for cloud computing services, and the easiest to relocate. This two pronged approach will ensure DOL addresses the business challenges and accelerate our overall cloud-smart migration.

Recommendation 22: *The Secretary of Labor should ensure that the CIO of Labor establishes a consistent and repeatable mechanism to track savings and cost avoidances from the migration and deployment of cloud services.*

DOL Response: DOL concurs with the GAO recommendation. DOL has developed a baseline to determine the cost of its "on premises" systems. This baseline was developed to track the recurring monthly costs of the "virtual assets" within the DOL Common Operating Environment. DOL will use this baseline to assess the cost savings for systems and services that have transitioned to the cloud. Savings tracked include both direct cost savings and cost avoidance.

Sincerely,

Bryan Slater
Assistant Secretary for
Administration and Management

APPENDIX XVII: COMMENTS FROM THE SMALL BUSINESS ADMINISTRATION

U.S. Small Business Administration

February 26, 2019

Ms. Carol C. Harris
Director, Information Technology Acquisition Management Issues
U.S. Government Accountability Office
441 G Street, N.W.
Washington, DC 20548

Dear Ms. Harris:

Thank you for providing the U. S. Small Business Administration (SBA) with a copy of the Government Accountability Office (GAO) draft report titled "Cloud Computing: Agencies Have Increased Usage and Realized Benefits, but Cost and Savings Data Need to be Better Tracked", GAO-19-58 (102299). The draft report discusses selected agencies' progress in implementing cloud services and the extent to which they increased spending on cloud services and achieved cost savings or cost avoidances. SBA is pleased that GAO reported SBA's success in reinvesting cost savings from consolidating data centers to the cloud into the agency's cloud modernization efforts with no additional budgeted funding.

SBA has reviewed the draft report and agrees with the one recommendation issued to the agency by GAO.

Recommendation 32: The Administrator of the Small Business Administration should ensure that the CIO of SBA establishes a consistent and repeatable mechanism to track savings and cost avoidances from the migration and deployment of cloud services.

SBA Response: Concur. SBA Office of the Chief Information Officer (OCIO), and the appropriate program offices will establish a consistent and repeatable process to track savings and cost avoidances from the migration and deployment of cloud services. We anticipate establishment of this process by September 30, 2019.

Thank you for the opportunity to comment on this draft report. Technical comments will be provided under separate cover. SBA appreciates GAO's consideration of our comments prior to publishing the final report.

Sincerely,

MARIA ROAT Digitally signed by MARIA ROAT Date: 2019.02.26 12:46:38 -05'00'

Maria Roat
Chief Information Officer

APPENDIX XVIII: COMMENTS FROM THE SOCIAL SECURITY ADMINISTRATION

SOCIAL SECURITY

Office of the Commissioner

January 11, 2019

Ms. Carol C. Harris
Director, Information Technology
Acquisition Management Issues
United States Government Accountability Office
441 G Street, NW
Washington, DC 20548

Dear Ms. Harris,

Thank you for the opportunity to review the draft report, "CLOUD COMPUTING: Agencies Have Increased Usage and Realized Benefits, But Cost and Savings Data Need to be Better Tracked" (GAO-19-58). Please see our enclosed comments.

Sincerely,

Stephanie Hall

Stephanie Hall
Acting Deputy Chief of Staff

SSA COMMENTS ON THE OFFICE OF THE GOVERNMENT ACCOUNTABILITY OFFICE (GAO) DRAFT REPORT, "CLOUD COMPUTING: AGENCIES HAVE INCREASED USAGE AND REALIZED BENEFITS, BUT COST AND SAVINGS DATA NEED TO BE BETTER TRACKED" (GAO-19-58)

We have extended our Agency Cloud Initiative (ACI) data center capability to include cloud infrastructure options for service delivery. We are executing an ACI hybrid strategy to benefit public services and employee functions across service channels, which include web, telephone, and in-person service. Our ACI supports Cloud Smart policy for using effective infrastructure platforms for new information technology investments as well as current workloads. We continue to monitor leading practices in the marketplace for applicability to agency operations, adapt enterprise governance methods and processes for operations, and invest in learning opportunities for relevant resource skills. Although we have been unable to quantify budgetary savings, we will continue to look for ways to measure cost savings that we may redirect to other Cloud or modernization activities. We will also adhere to any guidance OMB issues regarding the reporting of savings and cost avoidances related to the migration and deployment of cloud services.

Our responses to the recommendations are below. We also provided technical comments at the staff level for GAO's consideration.

SSA's Recommendation 1 – GAO's Recommendation 33

Ensure that the CIO of SSA updates the agency's guidance on assessing IT investments for suitability for cloud computing services to include a requirement to assess existing investments for these services.

Response

We agree.

SSA's Recommendation 2 – GAO's Recommendation 34

Ensure that the CIO of SSA completes an assessment of all IT investments for suitability for migration to a cloud computing service, in accordance with OMB guidance.

Response

We agree.

SSA's Recommendation 3 – GAO's Recommendation 35

Ensure that the CIO of SSA establishes a consistent and repeatable mechanism to track savings and cost avoidances from the migration and deployment of cloud services.

Response

We agree.

APPENDIX XIX: COMMENTS FROM THE DEPARTMENT OF DEFENSE

DEPARTMENT OF DEFENSE
6000 DEFENSE PENTAGON
WASHINGTON D.C. 20301-6000

CHIEF INFORMATION OFFICER

FEB 27 2019

Ms. Carol C. Harris
Director, Information Technology
U.S. Government Accountability Office
441 G Street, NW, Washington, DC 20548

Dear Ms. Harris:

This is the Department of Defense (DoD) response to the Government Accounting Office (GAO) Draft Report, GAO-19-58, 'CLOUD COMPUTING: Agencies Have Increased Usage and Realized Benefits, but Cost and Savings Data Need to be Better Tracked,' dated December 13, 2018 (GAO Code 102299).

RECOMMENDATION 1: The GAO recommends that the Secretary of Defense should ensure that the Chief Information Officer of Defense completes an assessment of all Information Technology (IT) investments for suitability for migration to a cloud computing service, in accordance with Office of Management and Budget (OMB) guidance.

DoD RESPONSE: Concur

In Q4FY18, the Department completed an assessment of all cloud-related efforts and investments across DoD to establish a cloud baseline. Additionally, DoD CIO performed assessments of Fourth Estate IT investments between Q3FY18 and Q4FY18 to identify cloud computing service candidates and is working with Fourth Estate IT investment owners to develop cloud migration plans based on those assessments. By the end of Q1FY20, DoD CIO and the Chief Management Officer (CMO) will jointly publish guidance and policy requiring DoD Components to conduct rationalization of their business and IT applications in alignment with the Department's enterprise-wide process for conducting software application rationalization and the DoD Cloud Strategy.

RECOMMENDATION 2: The GAO recommends that the Secretary of Defense should ensure that the CIO of Defense establishes a consistent and repeatable mechanism to track savings and cost avoidances from the migration and deployment of cloud services.

DoD RESPONSE: Non-concur

The Department has established a methodology for capturing funds requested for all cloud computing services procured by the Department and funds requested to migrate to a cloud computing environment. DoD's cloud computing services budget exhibit, which uses currently available DoD IT budget material, includes details regarding use of commercial and in-house cloud computing services by DoD Components. This exhibit will, for fiscal year 2021, expand to include costs associated with cloud migration. This effort supports the GAO finding that, "having complete data on spending for cloud services is critical to ensure that agencies can provide effective management and oversight of their cloud use, and that OMB and lawmakers can hold CIOs accountable for the performance of these cloud investments."

The Department does not currently plan to capture savings and cost avoidance associated with migration or deployment of cloud services since there is currently no standard, consistent way to capture such savings or cost avoidance. The Department will work with OMB on whether and how to collect such information, and, if practical, report such information in accordance with OMB guidance. The Department believes that although cost savings and cost avoidance may be an outcome of migrating to cloud, it is not the major decision factor for migrating or deploying cloud services in DoD. The Department is leveraging cloud technology and services primarily for its technical benefits, not for its potential costs savings and/or cost avoidance. The Department's recently released cloud strategy emphasizes this by identifying the following strategic challenges that will be addressed by DoD's multi-cloud, multi-vendor strategy:

- Enable Exponential Growth
- Scale for the Episodic Nature of the DoD Mission
- Proactively Address Cyber Challenges
- Enable Artificial Intelligence and Data Transparency
- Extend Tactical Support to the Warfighter at the Edge
- Take Advantage of Resiliency in the Cloud
- Drive IT Reform at DoD

Sincerely,

Dana Deasy

In: New Strategies in Cloud Computing ISBN: 978-1-53616-907-2
Editor: Luc Larivière

Chapter 2

FROM CLOUD FIRST TO CLOUD SMART*

Suzette Kent

In the Report to the President on Federal IT Modernization, released publicly in 2017 in accordance with Executive Order 13800,[1] the Office of Management and Budget (OMB) pledged to update the Government's legacy Federal Cloud Computing Strategy ("Cloud First"). Fulfilling this promise, the Administration has developed a new strategy to accelerate agency adoption of cloud-based solutions: Cloud Smart.

Developed nearly a decade after its predecessor, Cloud Smart equips agencies with actionable information and recommendations gleaned from some of the country's most impactful public and private sector use cases.[2]

* This is an edited, reformatted and augmented version of an official website of the United States Government Publication, Office of Management and Budget, Office of the Federal Chief Information Officer.

[1] Executive Order 13800, Strengthening the Cybersecurity of Federal Networks and Critical Infrastructure.

[2] Report to the President on Federal IT Modernization.

Beyond Cloud First, which granted agencies broad authority to adopt cloud-based solutions, Cloud Smart offers practical implementation guidance for Government missions to fully actualize the promise and potential of cloud-based technologies while ensuring thoughtful execution that incorporates practical realities.

The new strategy is founded on three key pillars of successful cloud adoption: security, procurement, and workforce. Collectively, these elements embody the interdisciplinary approach to IT modernization that the Federal enterprise needs in order to provide improved return on its investments, enhanced security, and higher quality services to the American people.

KEY ACTIONS

The Chief Information Officers Council (CIO Council) has developed a list of action items to execute the Cloud Smart strategy. These actions will constitute a work plan aimed at creating and updating programs, policies, and resources that the whole of Government will use to advance the Cloud Smart agenda.

Additionally, all Federal agencies will rationalize their application portfolios to drive Federal cloud adoption. The rationalization process will involve reducing an application portfolio by 1) assessing the need for and usage of applications; and 2) discarding obsolete, redundant, or overly resource-intensive applications. Decreased application management responsibilities will free agencies to focus on improving service delivery by optimizing their remaining applications.

To support these rationalization efforts, the CIO Council will develop best practices and other resources. Furthermore, while the initial Cloud Smart work plan will be executed over an eighteen-month period, its actions will be refreshed continuously as needed to keep up with the changing cloud market and emerging technologies.

Cloud at a Glance

Redefining Cloud Computing

The term "cloud" is often used broadly in the Federal Government for any technology solution provided by an outside vendor. The National Institute of Standards and Technology (NIST) defined several cloud deployment models as progressive increases in management by vendors, from Infrastructure as a Service (IaaS) where vendors provide the infrastructure and hardware, to Platform as a Service (PaaS) where vendors provide a managed environment for a customer's application, to Software as a Service (SaaS) where vendors provide a fully managed application and customers need only supply their data. In practice, many major vendor offerings no longer have such well-defined boundaries. Notwithstanding the term's common usage, the term "cloud" is most accurately applied to those solutions that exhibit five essential characteristics of cloud computing, as defined by NIST: on-demand service, broad network access, resource pooling, rapid elasticity, and measured service.[3]

These characteristics and the solutions that exhibit them are provider-agnostic – meaning anyone can develop and deploy a cloud solution, whether an outside vendor or a Federal agency. Industry has moved to a more finely differentiated set of capabilities offered at different system layers, making possible nearly any combination of various components managed by either a vendor, a Government agency, or a mix of both. Industries that are leading in technology innovation have also demonstrated that hybrid and multi-cloud environments can be effective and efficient for managing workloads. As a result, the Cloud Smart Strategy encourages agencies to think of cloud as an array of solutions that offer many capabilities and management options to enhance mission and service delivery.

Furthermore, Cloud Smart operates on the principle that agencies should be equipped to evaluate their options based on their service and

---

[3] NIST. "The NIST Definition of Cloud Computing." Special Publication 800-145.

mission needs, technical requirements, and existing policy limitations. Computing and technology decisions should also consider customer impact balanced against cost and cybersecurity risk management criteria. Additionally, agencies need to weigh the long-term inefficiencies of migrating applications as-is into cloud environments against the immediate financial costs of modernizing in advance or replacing them altogether.

Cloud adoption strategies that successfully meet the intent of Cloud Smart should not be developed around the question of who owns which resources or what anticipated cost savings exist. Instead, agencies should assess their requirements and seek the environments and solutions, cloud or otherwise, that best enable them to achieve their mission goals while being good stewards of taxpayer resources.

Modernization and Maturity

To realize the full benefit of cloud technology, agencies must cultivate an organizational mindset of constant improvement and learning. Modernization is not a commitment that is sustained solely by interventions once every decade. Rather, modernization is a constant state of change and part of the day-to-day business of technology at every agency. Critical to fostering this mindset of constant improvement is agency leadership's prioritization of the training and education of their staff, detailed and comprehensive migration planning, and a focus on balancing solution sustainability with the incorporation of new capabilities into agency operating environments. To that end, agencies will need to iteratively improve policies, technical guidance, and business requirements to match changing needs, drive positive outcomes, and prevent their IT portfolio from becoming obsolete.

Agencies should conduct regular evaluations of customer experience and user needs to ensure that their solutions successfully foster efficiency, accessibility, and privacy.[4] Additionally, agencies should regularly

[4] See Circular A-11, Section 280 for a discussion of customer experience requirements for agencies.

rationalize and update their applications, migrating as needed, to reduce the risk of large-scale failure, better allocate their resources, and provide staff with adequate time to become familiar with contemporary product management techniques. Agencies must also track their growth in areas where decisions about technology intersect other disciplines. Namely, serious consideration and investment should be dedicated to the three key pillars of successful cloud adoption: security, procurement, and workforce.

Given the distributed nature of cloud and the growing number of discrete capabilities and deployment models available to choose from, agencies might consider moving or adding security and privacy controls to the data layer itself, rather than just where they have historically resided at the network perimeter. By doing so, agencies can improve their overall security and privacy posture, empowering them to fully embrace cloud technologies while granting them peace of mind that the confidentiality and integrity of their data are intact.

To realize not only the security benefits of cloud infrastructure, but also its benefits related to scalability and speed-to-market, agencies should utilize mature agile development practices, including DevSecOps. The use of automated and assistive technologies such as artificial intelligence and machine learning can help agencies to further improve security. Agencies should also review their IT portfolios regularly to determine modernization plans for existing tools and compare potential service offerings designated as Best In Class (BIC) solutions for maximized return on investment.

Furthermore, providing staff with training and other educational resources is essential to fostering maturity in the areas of privacy, security, and procurement. Agency IT staff should become familiar with lean product management, agile development, continuous delivery, and automated infrastructure at the team and program level as part of any modernization plan. Additionally, non-IT staff supporting privacy, security, and procurement should receive training in the multiple core disciplines outlined above. Sustained progress in these areas of staff training is foundational to the successful implementation of new cloud efforts.

Consistent with the requirements of the Federal Information Technology Acquisition Reform Act,[5] the agency Chief Information Officer (CIO) should oversee modernization processes to help find opportunities for enterprise-wide improvement. Additional involvement of the Chief Financial Officer can help properly budget for planning, evaluation, and technology adoption. CIOs should also incorporate feedback from business units and end users affected by modernization projects to minimize disruption to mission delivery.

SECURITY

Agencies should take a risk-based approach to securing cloud environments. As recommended by the Report to the President on Federal IT Modernization, agencies should emphasize “data-level protections and fully leverage modern virtualized technologies.”[6] This requires that agencies place an emphasis on protections at the data layer in addition to the network and physical infrastructure layers, transitioning to a multi-layer defense strategy, otherwise known as defense-in-depth.

Critical to the success of this security strategy in the context of Cloud Smart is the assurance of confidentiality, integrity, and availability of Federal information as it traverses networks and rests within systems, regardless of whether those environments are managed locally, off-premises, by a Government entity, or by a contractor. Additionally, it is essential that agencies perform continuous monitoring to detect malicious activity and dedicate effort to improving systems governance.

Successfully managing cloud adoption risks requires collaboration between agency leadership, mission owners, technology practitioners, and governance bodies. Coordination between information security and privacy programs is necessary to ensure compliance with applicable privacy requirements and for the successful identification and management of risks

---

[5] 40 U.S.C. § 11319.

[6] https://itmodernization.cio.gov/.

to individuals when processing personally identifiable information (PII).[7] Senior Agency Officials for Privacy (SAOPs)[8] are responsible for managing the risk that may result from the creation, collection, use, and retention of PII, and have an important role to play when making decisions about the adoption of technology and processes that concern or impact the management of PII.

Cloud Smart encourages agencies to approach security and privacy in terms of intended outcomes and capabilities. The following programs are major elements of the Federal security strategy that must evolve alongside technological progress to allow agencies to take such a holistic and outcome-driven approach.

Trusted Internet Connections

Released in 2007, OMB Memorandum M-08-05[9] established new requirements for Federal agencies with the intent to reduce the Federal-wide number of external network connections while standardizing their security. Since the policy's release, agency network traffic in compliance with OMB requirements has flown exclusively through a limited number of external connections, known as Trusted Internet Connections (TICs). While this initial architectural concept served an important purpose at its inception, at a time when networking was constrained by physical limitations and agency approaches to network security were not standardized and highly fragmented, the technology landscape has evolved to provide agencies with more tools, technologies, and approaches to secure their data, leaving the once-useful TIC construct now relatively inflexible and incompatible with many agencies' requirements. With the proliferation of private-sector cloud offerings, the emergence of software-defined networks, and an increasingly mobile workforce, the TIC model

[7] As defined in Circular A-130, the term "personally identifiable information" means [...].
[8] M-16-24, Role and Designation of Senior Agency Officials for Privacy.
[9] M-08-05 Implementation of Trusted Internet Connections (TIC).

must compete with newer, more flexible solutions that provide equal or greater security, or it must evolve as well.

Electing the latter option, the Department of Homeland Security (DHS) is working with various agencies to pilot agency-specific approaches that meet the objectives and intent of M-08-05 while minimizing technical constraints posed by the policy's one-size-fits-all TIC model. These newer, less rigid approaches will be incorporated into updated TIC Reference Architectures to highlight use cases wherein security objectives can be met without routing all traffic through a prescribed set of physical access points. The TIC Reference Architectures will also demonstrate how different use cases that do not require traffic to be routed through a TIC can address the requirements for government-wide intrusion detection and prevention efforts, such as the EINSTEIN Program while also incorporating DHS-designated controls, which have been designed to ensure a baseline level of security across the Federal enterprise.[10]

By taking these actions to expand the options available to agencies to secure their networks and data, the collective ability of the Federal Government to take advantage of new paradigms, such as zero trust networks, is heightened, as its effectiveness in managing risk.

Continuous Data Protection and Awareness

Given the ease of flow of federally-owned data from internal networks to external, end-user devices, encryption and modern Identity, Credential, and Access Management (ICAM) implementation is essential. Encryption and ICAM implementation is particularly relevant in the context of cloud-based environments, namely in those instances where an agency is partnering with an external service provider to manage network visibility and data protection.

---

[10] https://www.dhs.gov/einstein.

To ensure continuity of information security during and after the migration process, it is incumbent upon agencies to thoroughly assess their operational, policy, and business requirements and advocate for themselves when brokering new arrangements with cloud service providers. Regardless of provider type – commercial or Federal – agencies should consider having agreements in place with all providers regarding access to, and use of, log data given its importance in effectively conducting information security operations. Moreover, as each agency is the custodian of its information on behalf of the public, each agency is responsible for determining its own governance model for cloud-hosted data that aligns with its identity and credential management systems. To that end, where a cloud solution is deployed by a vendor, a Service Level Agreement (SLA) should be in place that provides the agency with continuous awareness of the confidentiality, integrity, and availability of its information.

Furthermore, agencies should be made aware if their information will reside on a third-party information system prior to signing any service agreement; agencies should be provided continuous access to log data; and must be notified promptly if a cybersecurity incident, breach,[11] or other adverse event occurs or is suspected to have occurred that involves any information or information systems covered by a service agreement with a cloud service provider.

To further enable continuous data protection and awareness, agencies and their partners should regularly engage in reciprocal information sharing. Cybersecurity requires public-private collaboration, and as more Federal entities adopt commercial cloud solutions, customers and providers should work together to protect information. Recognizing the criticality of tools and analytical capabilities that scale across multi-cloud environments in the facilitation of continuous visibility and information sharing, DHS's Continuous Diagnostics and Mitigation (CDM) Program[12] also continues

---

[11] As defined in OMB M-17-12, a breach is defined as "The loss of control, compromise, unauthorized disclosure, unauthorized acquisition, or any similar occurrence where (1) a person other than an authorized user accesses or potentially accesses personally identifiable information or (2) an authorized user accesses or potentially accesses personally identifiable information for an other than authorized purpose."

[12] https://www.dhs.gov/cdm.

to evolve so that agencies are equipped with the monitoring capabilities they need to understand their cyber risk in the cloud.

FedRAMP

The Federal Risk and Authorization Management Program (FedRAMP) provides a standardized government-wide approach to security assessment, authorization, and continuous monitoring of cloud services. Offering cloud service providers the opportunity to demonstrate their ability to meet Federal security requirements through standardized baselines has allowed for a flourishing marketplace of vetted providers to develop. It has also allowed agencies to adapt from arcane legacy technology to mission-centric and cost-effective cloud-based systems in a more rapid, consistent, and secure manner.

Although the FedRAMP program management office has drastically reduced the amount of time it takes to authorize a cloud service provider, there is still work to be done to address the underlying issues that contribute to the relatively slow pace of assessment. For example, a lack of reciprocity across agencies when adopting FedRAMP authorizations has led to significant duplication of effort when assessing security for product deployment. In addition, a large number of agency-specific processes has made it complicated for agencies to issue an Authorization to Operate (ATO) for solutions, even when using existing authorized cloud service providers. In fact, despite the reiterated importance of enterprise risk management,[13] agencies continue to cite major obstacles with their own policies and practices.

To tackle these challenges, several initiatives aimed at overall process evolution as well as strategies for accelerating common ATO agreements are under development. These efforts intend to drive better and more automated control inheritance and monitoring, a prioritized approach to control implementation, and more normalized control use across the

[13] As established through Circular A-130, the Federal Information Processing Standards, and NIST Special Publications.

Federal enterprise. Advancements to cloud ATO development will be used to inform overall ATO reform, which will involve the revision of NIST special publications. This will also reestablish FedRAMP's role in the risk assessment process as a verification check for agencies as they make informed decisions about the cloud solutions that they deploy, rather than a panacea for all matters related to the risk associated with any implementation of a cloud solution. OMB and GSA will continue to promote alignment and reuse of ATO determinations and closely examine agency-identified obstacles in that effort.

Finally, enhancing the skillsets of the Federal workforce around cloud security in the FedRAMP program will allow the Federal Government to continue to increase the efficiency and effectiveness of agency security practices in adopting cloud systems, while reducing the burden on security professionals, providers, and agency leadership. Assembling a cadre of professionals and providing direct engagement with all aspects of the security authorization process will build a common and comprehensive understanding of cloud security and enable more trust when sharing ATOs. Agencies are also encouraged to take a multidisciplinary approach to hiring and training their workforce, as well as provide community spaces where digital services experts, information security professionals, procurement specialists, and others with a mutual interest in effective, secure cloud adoption can collaborate on current challenges and opportunities in the cloud computing space.

PROCUREMENT

Industry partners, interagency working groups, and individual agencies have provided the Federal IT and acquisition communities with a wide selection of recommended actions to accelerate the adoption of cloud solutions. Some of these recommendations have been translated into Federal-wide guidance, but there remains a lack of consistency across agency implementations and information sharing on best practices. In the absence of comprehensive guidance, agencies must search across multiple sources to gain a basic understanding of the various types of cloud services sold in the commercial marketplace, the different offerings available on existing government-wide contracts, and the best way to evaluate which approach is best for a given requirement.

As a result of ubiquitous private sector use of cloud computing, agencies often purchase services through contracts that, while not specifically marketed as a cloud-based service or capability, involve placement of agency information into cloud-based systems for processing or storage. This trend creates potential security and privacy concerns and warrants that agencies pay greater attention to professional services contracts, where the storage of Federal information on non-Government information systems may be implied.

To address these challenges, agencies will need to use a variety of approaches that leverage the strength of Federal Government's bulk purchasing power, the shared knowledge of good acquisition principles, as well as relevant risk management practices. As part of the Cloud Smart multidisciplinary approach, agencies will also need to place security and privacy considerations at the forefront of any procurement effort, and to avoid vendor lock-in, they should evaluate the business process dependencies of any new solution. They should also update their business continuity and disaster recovery plans to include contingencies involving the sudden interruption or termination of service. Detailed below are additional considerations for Federal procurement professionals navigating the IT space and IT professionals seeking guidance on common practices to ensure the cost-effective, safeguarded procurement of cloud-based solutions.

Category Management

Category management describes the strategic business practice that the Federal Government is implementing to buy smarter and more like a single enterprise. Specifically, government-wide category management aims to: 1) deliver more savings, value, and efficiency for Federal agencies; 2) eliminate unnecessary contract redundancies; and 3) meet the Government's small business goals.

Category management simplifies the process for industry to do business with the Government by reducing duplicative contracts and decreasing costs for bids, proposals, and contract administration. It also offers the Federal Government the opportunity to improve buying practices that support Cloud Smart strategies, increase adoption of proven cloud vehicles in the Federal marketplace, and develop new vehicles to address emerging demands. OMB has published a memorandum[14] that outlines guidance on the implementation of category management principles and use of common contract solutions.

Service Level Agreements

A Service Level Agreement (SLA) between a customer and a service provider defines the level of performance expected from a service provider, how that performance will be measured, and what enforcement mechanisms will be used to ensure the specified levels are achieved. While not standardized in the Federal Acquisition Regulation (FAR),[15] SLAs are incorporated through contract clauses and quality assurance contractual provisions. In legacy technology environments, these agreements represent a critical element of negotiation with suppliers. Unfortunately, the term "Service Level Agreement" itself has become overloaded with multiple meanings depending on context, which has resulted in increased uncertainty around the best approach to achieve better outcomes for agencies. Cloud Smart offers a two-track approach to smarter cloud purchasing and usage across Federal agencies through improvements to SLA use, as outlined below.

First, it is important to note that candidates for inclusion as standard clauses that apply to commercial items in the FAR – including new SLAs –

---

[14] M-19-13, Category Management: Making Smarter Use of Common Contract Solutions and Practices.

[15] 48 C.F.R. § 12.301.

must generally meet at least one of the following criteria: (1) the clause is required to implement a provision of law applicable to the acquisition of commercial items; or (2) the clause is generally consistent with customary commercial practice. Keeping these stipulations in mind, the first track of activities to support the effective use of SLAs involve the government-wide review and selection of contractual terms and conditions specific to cloud-based commercial offerings that are good candidates for standardized use across agencies. Standardizing cloud contract SLAs will provide more effective, efficient, and secure cloud procurement outcomes for agencies, while also enabling enhanced management of risk across the Federal enterprise through greater consistency and transparency in negotiations with commercial suppliers.

Second, heads of executive agencies are accountable for managing the risk to their enterprises, and that responsibility remains with the agency head even with respect to contractor-operated systems. An important element of acquiring cloud services is clarity in what services a cloud provider performs and at what level. Therefore, to facilitate effective risk management by way of their relationships with commercial cloud service providers, agencies should granularly articulate roles and responsibilities, establish clear performance metrics, and implement remediation plans for non-compliance. Such governance, architecture, and operational clarity would help ensure that services are performed as intended, and, when paired with the right SLA language, would offer agencies a way to mitigate risk while optimizing the performance and efficiency of their newly procured cloud-based solution.

Security Requirements for Contracts

It is essential that agencies consider and manage security and privacy risk to information and mission services when making cloud procurement and deployment decisions. Incorporating this approach as part of the acquisitions process and system development lifecycle, the Federal CIO has published an updated High Value Asset (HVA) memorandum[16] that builds on the previous initiative and adds considerations for managing risk across hybrid environments. Agencies must now ensure that contracts impacting their HVAs, including those managed and operated in the cloud, include requirements that provide agencies with continuous visibility of the asset. The guidance enables agencies to proactively assess the security and privacy posture of their assets and seeks to increase HVA trustworthiness by requiring developers, manufacturers, and vendors to employ best practices for designing, deploying, and securing systems. This will drive a targeted integration of security and privacy design principles, secure coding techniques, and trusted computing methods.

WORKFORCE

The Federal IT workforce plays an integral role in the execution of agency missions, delivery of services to the public, and provision of security to the nation's essential systems and information. In the same way that agencies cannot outsource risk to commercial suppliers, neither can they outsource decision-making. Agencies' immediate and sustained investment in the Federal workforce is critical to the enhanced quality, security, and impact of services delivered to taxpayers. Without it, improvements to the Federal Government's technology infrastructure, and the successful proliferation of the Cloud Smart strategy will not be fully realized.

As agencies adopt and migrate to cloud platforms, the impact that these migrations have on the Federal workforce needs to be examined. Specifically, agencies should identify potential skills gaps that emerge as a result of a transition to cloud-based services, and, where needed, equip their existing staff with additional skills and knowledge to keep up with the ever-expanding list of technology options available to procure and deploy.

---

[16] M-19-03 Enhancing the High Value Asset Program.

Agencies should not restrict themselves to a single model for workforce transformation. Strategies might include development programs for emerging talent or future leaders; apprenticeship programs; initiatives to convert non-IT personnel where aptitudes align; or exchanges of qualified personnel through public-private partnerships or interagency detail opportunities. Generally, agencies' cloud strategies and policies should also include a workforce development and planning component that tailors a transformation and training approach to that agency. Additionally, in the event that an impact to the existing workforce has been identified, this approach should include a cross-walk of new skills and occupational categories with legacy occupational categories to foster clarity and ease of transition.

Identifying Skill Gaps for Current and Future Work Roles

Successful adoption of cloud solutions requires a workforce that understands how to manage the complexities of a migration as well as how to support a cloud environment once fully deployed. Agency CIOs, Chief Human Capital Officers (CHCOs), and SAOPs should collaboratively conduct a skills gap analysis that maps current IT workforce resources to future skill and position requirements. Agencies are strongly encouraged to leverage industry projections to help predict future workforce skill and position requirements, especially for IT roles.

The National Initiative for Cybersecurity Education (NICE) Cybersecurity Workforce Framework, whose use by Federal agencies is mandated in the Federal Cybersecurity Workforce Assessment Act of 2015,[17] recommends an approach for identifying all Federal civilian positions performing IT or cybersecurity-related functions.[18] While Federal agencies should continue to comply with the NICE Framework to help standardize government-wide cybersecurity workforce gap assessments, they are encouraged to conduct their own enterprise-wide skills gap

[17] https://www.congress.gov/bill/114th-congress/house-bill/2029/text

[18] NIST Special Publication 800-181 – NICE Cybersecurity Workforce Framework.

analysis to ensure inclusion of all current and future IT skills and positions specific to their workforce requirements.

Reskilling and Retaining Current Federal Employees

Current employees may lack the skills or knowledge required to facilitate a cloud migration or to maintain the environment once migrated. Therefore, agencies must conduct their skills gap analysis to identify both technical and non-technical skill and position gaps so that leadership may determine which deficiencies are the most significant and/or represent a critical mission need.

Agencies should address the most pressing deficits found by a workforce gap analysis by developing employee reskilling strategies that focus on training and professional development opportunities, such as OMB's Digital Information Technology Acquisition Professional (DITAP) program.[19] A complete reskilling strategy will account for technical as well as non-technical needs. Given that cloud computing is relatively new territory for acquisition professionals, such as Chief Acquisition Officers, Contracting Officers, and Project Managers, should take advantage of existing guidance from resources such as the OMB, the Federal Acquisition Institute,[20] and the TechFAR HUB.[21] Consistent with guidance for IT acquisition roles, agencies may also benefit from developing a specialized team or expanding the use of IT acquisition cadres.[22]

Lastly, CHCOs and Chief Learning Officers should help determine optimal training and redeployment options (e.g., certification and rotational programs) to utilize reskilled employees. As with many new technology initiatives, agencies should expect an aggressive initial training period and should also plan for ongoing education and experimentation in this rapidly evolving field. Providing valuable employees access to continuing

---

[19] https://techfarhub.cio.gov/initiatives/ditap/.

[20] https://www.fai.gov/.

[21] https://techfarhub.cio.gov/.

[22] https://www.whitehouse.gov/sites/whitehouse.gov/files/omb/procurement/memo/guidance-for-specialized-acquisition-cadres.pdf .

education resources and opportunities to apply their knowledge will both inform new cloud initiatives and promote job satisfaction.

Above all else, the success of initiatives like these is dependent on the support of champions in executive leadership who broadly vocalize their backing of the effort and who remove roadblocks that discourage or prevent employees from pursuing reskilling or certification opportunities. While finding the right champion presents its own challenges, the ability of this person to amplify the reach and results of the agency's initiative is invaluable.

Recruiting and Hiring to Address Skill Gaps

The U.S. Department of Labor's Bureau of Labor Statistics reports that cloud computing is a major factor in technology occupation growth, which is projected to expand 13% from 2016 to 2026.[23] In addition to reskilling current employees to address the most critical skill and position gaps, agencies should continuously evaluate and update their recruitment and hiring strategies. Key strategies include leveraging industry recruitment best practices, expanding the use of pay flexibilities, and removing bureaucratic barriers to hiring staff expeditiously. Agencies must build a pipeline to continuously feed cybersecurity and IT talent into the Federal Government.

The market for technology professionals with cloud computing skill sets is extremely competitive. When possible, agencies should leverage techniques used by the private sector to attract and hire the best candidates to the Federal Government. In coordination with their Chief Human Capital Officers, agencies should execute proactive recruitment and retention strategies such as:

---

[23] Bureau of Labor Statistics – *Occupational Outlook Handbook: Computer and Information Technology Occupations*, as of April 2018 - https://www.bls.gov/ooh/computer-and-information-technology/home.htm.

- Attending industry conferences with career fairs;
- Holding national hiring events to strengthen awareness and outreach;
- Developing "most wanted" talent advertisements to showcase critical needs;
- Ensuring that job postings on places like USAJOBS properly reflect needed skills;
- Engaging candidates through social media platforms;
- Profiling and sharing current employee experiences;
- Offering incentives to employees for relevant professional development;
- Supporting engagement of Federal employees in appropriate industry forums to increase visibility and access;
- Leveraging merit promotion hiring procedures to retain, promote, or redeploy current Federal employees when appropriate;
- Leveraging hiring and retention authorities approved by the Office of Personnel Management, such as incentive pay and awards;
- Leveraging Scholarship for Service and other academic partnership to build a pipeline for graduates into the Federal Government; and
- Showcasing diversity and inclusion initiatives.

While the Government invests in efforts to recruit existing talent to the Federal workforce, it should also build a talent pipeline to expand the pool of qualified applicants. Through the Workforce Council, Federal Privacy Council, and the Chief Information Officer Council initiatives, the Government will continue to develop partnerships with community colleges, apprenticeship programs, and four-year institutions, in addition to leveraging partnerships that already exist. Expanded consideration of reskilling and upskilling solutions should be included in agency transition strategies as well.

Employee Communication, Engagement, and Transition Strategies

As agencies implement the Cloud Smart strategy, they should execute communication plans that help employees understand the changes that will occur. For example, migration to the cloud may require decommissioning legacy systems that have been in use for many years. Employees may feel reluctant, especially if positions will be redefined, to learn to operate new systems in a cloud environment. Agencies can ease workforce concerns by clearly articulating how the current workforce will align once cloud adoption is complete. Socializing a technology roadmap to include systems that will be migrating to the cloud, either completely or partially, and an outline of the change management process to include reskilling opportunities is strongly recommended. Agencies should also feel comfortable leveraging vendors involved in cloud migration activities to provide or support training for current employees.

Removing Bureaucratic Barriers to Hiring Talent Expeditiously

The demand for technology professionals with cloud computing skills sets is at an all-time high. This means that attracting, recruiting, and retaining the right individuals will take an executable human capital strategy with a streamlined hiring process and non-traditional incentives. Agencies with aggressive hiring timelines and competitive offers leveraging pay and recruitment incentives will attract talent. It is imperative for agency leadership to identify and promptly address bureaucratic barriers that hinder agencies from hiring talent in an expeditious manner.

Agencies have broad authorities under Title 5 of the United States Code to hire top IT and cybersecurity talent, and to provide candidates with superior qualifications or who address critical skill gaps with pay flexibilities and incentives. Agencies are strongly encouraged to use available hiring authorities, recruitment incentives, and student loan repayment benefits to hire professionals with highly sought-after cloud computing skills.

It is incumbent upon Federal agencies to ensure that their current and future workforce is prepared to support Federal cloud environments. Agency cloud strategies and commensurate workforce investments should enable leaders to develop and empower the IT and cybersecurity workforce with the skills required to achieve cloud migration goals and support the latest technology that will improve delivery of critical citizen services.

CONCLUSION

To be Cloud Smart, agencies must consider how to use their current resources to maximize value: reskilling and retraining staff, enhancing security postures, and using best practices and shared knowledge in acquisitions. Cloud Smart is about equipping agencies with the tools and knowledge they need to make these decisions for themselves, rather than a one-size-fits-all approach.

By leveraging modern technologies and practices, agencies will be able to harness new capabilities and expand existing abilities to enable their mission and deliver services to the public faster. To make this shift, instead of "buy before build", agencies will need to move to "solve before buy," addressing their service needs, fundamental requirements, and gaps in processes and skillsets before starting on a new procurement. By rationalizing their application portfolios regularly, agencies can continue to make modernization progress while targets move with the ever-changing technology landscape.

Since the release of the original draft of this strategy, OMB has worked with its partners – both Government agencies and in the private sector - to update policy guidance and create new resources for aiding cloud adoption. Yet there will always be more work to be done, as technology does not have a finish line. Embracing change as a core business principle is critical to delivering the best return on investment to the American people.

In: New Strategies in Cloud Computing ISBN: 978-1-53616-907-2
Editor: Luc Larivière

Chapter 3

DoD Cloud Strategy*

Department of Defense

Foreword

The Department of Defense (DoD) has entered the modern age ofwarfighting where the battlefield exists as much in the digital world as it does in the physical. Data and our ability to process data at the ready are differentiators to ensure mission success. Cloud is a fundamental component of the global infrastructure that will empower the warfighter with data and is critical to maintaining our military's technological advantage.

The DoD Cloud Strategy reasserts our commitment to cloud and the need to view cloud initiatives from an enterprise perspective for more effective adoption. It recognizes our experience over the past five years and identifies seven strategic objectives along with guiding principles to set a path forward. It emphasizes mission and tactical edge needs along

* This is an edited, reformatted and augmented version of Department of Defense Publication, dated December 2018.

with the requirement to prepare for artificial intelligence while accounting for protection and efficiencies.

The strategy drives implementation toward the enterprise cloud environment, an ecosystem composed of a General Purpose and Fit For Purpose clouds. It focuses implementation activities on two fundamental types of work: first is the stand up of cloud platforms ready to receive data and applications, and second is the ongoing work to migrate existing applications and to develop new applications in the cloud.

This effort is a Department priority. As we execute this strategy, we will continue to seek the active participation and commitment of all DoD Components to realize the benefits of cloud as we operate on the 21st century battlefield.

STRATEGIC ENVIRONMENT

> "If we fail to adapt ... at the speed of relevance, then our military forces ... will lose the very technical and tactical advantages we've enjoyed since World War II"
>
> - Secretary of Defense James N Mattis

Information is vital to United States (U.S.) national security and our ability to understand emerging threats, project power globally, conduct operations, support diplomatic efforts, and enable global economic viability. The Department of Defense (DoD) has multiple disjointed and stove-piped information systems distributed across modem and legacy infrastructure around the globe leading to a litany of problems that impact warfighters', decision makers', and DoD staff's ability to organize, analyze, secure, scale, and ultimately, capitalize on critical information to make timely, data-driven decisions. Today, the Department is largely constrained by physical resources, manpower limitations, organic skillsets and, oftentimes, laborious contracting processes to procure or grow storage and computing capabilities. In addition, the cyberspace domain continues to be an increasingly contested environment. In order for the U.S. to keep its strategic advantage, warfighters and the force that support them need to be provided with the proper capabilities and technologies to succeed.

To this end, commercial industry has made significant strides in addressing these challenges that the Department can leverage. Commercial cloud computing is a subscription-based service that provides network-based storage and compute resources. It allows users to store and access data and programs over the Internet rather than on a local computer hard drive. It also allows users to access information from anywhere at any time, effectively removing the need for the user to be in the same physical location as the hardware that stores the data. The Department must take full advantage of this technology enabler.

Inadequate Efficiency and Security in Information Technology

The physical information technology (IT) infrastructure deployed across the Department was purchased with a "maximum use" case in mind. In other words, the hardware can support the greatest possible expected demand, regardless of how infrequently that may happen, if at all. This causes much of the physical IT infrastructure purchased to be idle the majority of the time. Commercial cloud infrastructure works differently. It can scale dynamically to support resource demands with the benefit of only paying for the actual use. During most times, systems can be scaled down to support the minimal traffic that is the norm.

By owning and operating the physical hardware associated with on-premises data centers, the Department can incur unnecessary security risks and consume resources that could otherwise be realigned to support warfighters and the workforce in other mission areas. A combination of overly strict policies and procurement procedures make it difficult for DoD to ensure that both hardware and software are updated appropriately. The cost spent on physical hardware has a negative impact on diverting vital resources away from the warfighter and workforce as well as security impacts with a direct burden of responsibility for security updates of both hardware and software on DoD. The Department has historically been challenged to keep up with cyber threats to its IT infrastructure.

Disparate Cloud Efforts and Disjointed Implementation

DoD has not had clear guidance on cloud computing, adoption, and migration to provide unifying guidance or a coherent plan. This has made it difficult for the Department to embrace modern IT capabilities, to benefit from the efficiency and capacity offered by commercial cloud services, and to continue to evolve with technology at the speed of relevance.

The lack of guidance has led to Departmental inefficiencies and has hindered the Department in IT modernization efforts. It has led to disparate efforts with siloed teams, disjointed implementations with limited capability, siloed data, and inefficient acquisitions that cannot take advantage of economies of scale.

Lack of Cloud Fitness

DoD has stood up a number of clouds that have not been architected or designed for enterprise use. It is imperative that DoD has a cloud strategy to ensure that legacy applications are not moved to cloud without properly rearchitecting them to make use of the data, security, resiliency, and application advantages that cloud provides. Additionally, DoD should independently test and assess cloud network security to verify security compliance and incident response, and review all contractor and third-party testing results to ensure that perfonnance and security monitoring are sufficient. Systems that are not "cloud ready" will most likely use excessive amounts of cloud infrastructure resources, meaning they will not be any more efficient and will almost certainly cost more to operate. This is why system rationalization is critical, but, without proper guidance, many program offices may not do this properly or skip it entirely.

There is currently no enterprise-wide guidance on how to rationalize a system nor for assessing a system's cloud readiness. This gaphas further led to the siloeddata behavior mentioned earlier and restricted the ability of the Department to share infonnation. This enterprise-wide guidance is imperative to the success of any cloud strategy - enterprise or otherwise.

As DoD has continued to stand up independent clouds, we continue to dilute our already constrained cloud expertise. This enterprise cloud strategy provides the constructs for optimizing our use of cloud talent. To date, the DoD oftentimes relies on outside contracting finns to perfonn these assessments, never building the organic corporate knowledge to carry us into the future. Decisions are being made at high levels without in-house technical expertise as advisors. The Department must invest in its own future by building technical knowledge within.

Readiness for Artificial Intelligence (AI)

As DoD embarks to stand up a Joint Artificial Intelligence Center (JAIC), it will require an enterprise cloud infrastructure capability. An enterprise cloud will provide the common data and infrastructure platforms that will enable AI to meet the full promise of warfighter advantage.

DoD has created this DoD Cloud Strategy to align with the larger DoD cyber strategy, strengthening the security and resilience of the networks and systems that contribute to the Department's military advantage.

Strategic Objectives

DoD will continue to rely on its ability to process and disseminate information for military operations, intelligence collection, and related activities. To ensure this, the Department must address the unique mission requirements through a multi-cloud, multi-vendor strategy that incorporates a General Purpose cloud and Fit For Purpose clouds (reference Appendix A). To this end, this strategy will design objectives around solving these strategic challenges:

- Enable Exponential Growth
- Scale for the Episodic Nature of the DoD Mission
- Proactively Address Cyber Challenges
- Enable AI and Data Transparency
- Extend Tactical Support for the Warfighter at the Edge
- Take Advantage of Resiliency in the Cloud
- Drive IT Reform at DoD

Enable Exponential Growth

The pace of data growth is accelerating; in the last two years, the world produced 90% of all existing data. This is a trend that has been going on for a decade, with no end in sight; however, the Department's ability to access all of that data when and where it is needed has not evolved at the same pace. Modem computing capabilities can access, retrieve, manipulate, merge, analyze, and visualize data at machine speeds, providing substantial decision making advantages on the battlefield. To adapt to the continuously growing data environment, DoD requires an extensible and secure cloud environment that spans the homeland to the global tactical edge, as well as the ability to rapidly access computing and storage capacity to address warfighting challenges at the speed ofrelevance.

DoD relies on critical intelligence to make vital national security decisions. The quantity and quality of intelligence information has been the tipping point in numerous conflicts. As the quantity of raw information production increases, so does the struggle to organize, analyze, and distribute that information to make critical decisions.

DoD must continue to maintain its strategic advantage across the globe. In today's world, this cannot be done without laying the critical foundation needed to harness the power of its own data and information systems. This is the realization of cloud computing: the ability to organize, analyze, secure, scale, and ultimately, capitalize on critical information and fight in the digital age. These capabilities must be ubiquitous and available to all Department decision makers, warfighters, and staff.

Scale for the Episodic Nature of the DoD Mission (Elasticity)

By implementing a scalable solution, mission owners will gain significant efficiencies in the execution of mission capabilities and cyber operations by fully embracing the dynamic elasticity of commercial cloud architecture. The Department's cloud infrastructure will allow for provisioning and deprovisioning of resources automatically. This provides optimum asset utilization when compared to traditional IT infrastructure that is constantly in use, even when demand is minimal. This efficiency will also eventually improve the government's budgeting, billing, and payment practices by providing detailed resource usage reports for all mission owners. This transparency will further drive more efficiencies in the future on how applications are built.

Additionally, the cloud pay-for-use model will provide the flexibility to optimize costs across the IT portfolio and allow DoD to adapt to changing priorities, budgetary conditions, and industry developments. To achieve this cost transparency, strong governance will need to be put in place for how applications are built and data is transmitted and stored. As we develop these standards, implement them, and subsequently learn and better align our services and data to an enterprise solution, we can look to automated tools and techniques to better inform accurate tracking of financial execution of cloud resources.

Proactively Address Cyber Challenges

DoD must create a standard cloud-based cyber architecture that addresses the needs of commercial and internal-based clouds and encompasses infrastructure, applications, and data. This must include the ability to keep the environment "evergreen" in terms of security and technology.

DoD will produce a unified cybersecurity architecture that addresses cloud and the needs of classified and unclassified missions and data. The capabilities will be tested and assessed independently and frequently to ensure that cybersecurity attributes remain effective against developing threats.

DoD must embrace modern security mechanisms built into modern commercial cloud providers' platforms to ensure the security of these large amounts of data and to safeguard the information. This requires shifting the focus of security from the perimeter edge of the network to actively controlling use of the data itself. In addition to modern encryption algorithms and key management built into commercial cloud services, proper tagging of data will allow for it to be tracked and protected at the necessary levels. DoD will develop a Data Management Strategy that provides the focused discussion with respect to data.

In addition to DoD data security, each Cloud Service Provider will be integral to combating cyber challenges and securing the cloud. The Cloud Service Providers will automatically scan infrastructure resources and generated logs, which will be used to identify vulnerabilitiesearly and to make intrusion detection and mitigation in near-real time a reality across much of the enterprise. With the rise of hardware vulnerabilities, such as Spectre, and increased insider threat, a focus must be applied to both software and hardware-which change at an incredible pace. Keeping up with those changes is difficult, but failure to keep pace has created significant security risks and will only increase in the years to come. Here, again, modern commercial providers have addressed this problem. Moving infrastructure from DoD-managed, on-premises facilities to the cloud will take advantage of the rapid roll out of software and hardware updates.

Cloud Service Providers are able to shift workloads within their data centers such that updates are seamless to customers. Hardware with defects or vulnerabilities is constantly swapped out and software patches are applied with vigor in a secure and fault tolerant manner.

Although commercial cloud has many security advantages and opportunities for the Department, the transition to the commercial cloud environment also presents new security challenges. The transition from traditional IT management to the managed cloud service model alters the balance of visibility and control with ease of use, automation, leading edge technology adoption, and optimization of its information domain. The DoD CIO is responsible for defining the security guidelines in the cloud environment. The risk and the responsibility for executing the security in the cloud environment is shared between the Cloud Service Provider(s) and the system owners. DoD CIO will identify the command and control (C2) requirements of the shared cybersecurity responsibility model between DoD and commercial vendors to ensure standard execution of C2 responsibilities for DoD information in commercial cloud. The specific requirements of securing a cloud environment will strain the traditional technical workforce and requires specialized skills where the Department currently has limited expertise.

Historically, information security has been heavily focused on perimeter defense: limiting network access at the boundary. Unfortunately, this model is challenging for a commercial cloud environment where data is being accessed remotely and shared within and between deployments, regions, and from each Cloud Service Provider to other data locations, such as on-premises data centers at military installations. Therefore, the Department will shift its security focus from perimeter defense to securing data and services. This shift will be accomplished first through strong authentication for both people and machines and secure encryption mechanisms both at rest and in transit. In order to facilitate remote access, the DoD cloud environments will supply built- in cryptographic technology that enables organizations to encrypt communications by default. Since the information security responsibility is shared between the Department and its Cloud Service Providers, the Department will include language in all cloud computing contracts directing Cloud Service Providers to monitor their cloud infrastructure and maintain authenticated, encrypted logging of security-relevant events that generate an audit trail and are engineered to be resistant to tampering. To address the workforce strain in adopting these new security postures, the Department will include cloud adoption assistance and speciaUzed training for its workforce as a part of DoD Cloud Service Provider contracts.

Enable AI and Data Transparency

DoD must enable decision makers to use modern data analytics, such as AI and machine learning (ML), at the speed of relevance to make time-critical decisions rapidly in the field to support lethality and enhanced operational efficiency. The algorithms used to inform decisions are dependent on the Department's data and information being organized, secure, and visible in a common environment. An environment where data is stored in a multitude of disparate and disjointed stove pipes reduces the efficiency and tempo of the Department. To maximize the utility of cloud computing technologies, data must be managed properly and follow modern technologies like data lakes and data hubs, which are accelerated and amplified by cloud technology.

Data stored in an enterprise DoD cloud will be highly available, well-governed, and secure. Data will be the fuel that powers those advanced technologies, such as ML and AI. This critical decision making data will be made available through modern cloud networking, access control, and cross domain solutions to those who require access. Common data standards will be a key part of the Department's methodology for tagging, storing, accessing, and processing information. Ensuring an enterprise cloud environment will increase the transparency of this data, and drive the velocity of data analysis, processing, and decision making. Leveraging advances in commercial cloud security technologies will ensure the Department's information is protected at the appropriate level.

Commercial cloud provides the ability to scale and secure both the collection and the analysis of data stored in an enterprise DoD cloud. This gives mission owners the capability to make decisions with the most relevant information. The distributed nature of cloud computing allows for a more flexible execution environment while simultaneously providing increased information security. This allows for scaling and distributing data repository stores while maintaining security posture and providing new opportunities to obtain mission insights through data collaboration. Similarly, the computing power required for analysis of massive amounts of data can be scaled seamlessly in seconds. This ability to scale will ensure that mission execution is not hindered by insufficient computing and storage capacity and enable the creation of new information models that were previously unachievable.

Extend Tactical Support for the Warfighter at the Edge

The DoD cloud environment will serve mission O\Vners in every environment, across the range of military operations, from the tactical edge to the home front, both CONUS and OCONUS, and at all classification levels and disseminations (e.g., NOFORN and REL). We must embrace computing solutions that enable warfighters in their environment versus forcing them to conform to the current environment of siloed data and legacy applications. The integration and operation of computing solutions will be straightforward and repeatable, regardless of the required classification level of the system. This will allow warfighters to make data driven decisions and enhance DoD ability to share data with allies and operate as a coalition force. The security of the classified environments will support the level demanded by mission requirements.

Industry has made huge strides in disconnected operations. The Department's General Purpose and Fit For Purpose clouds will capitalize on these efforts to provide the warfighter with the latest technology where they need it and when they need it regardless of the environment. Cloud devices employed by warfighters at the tactical edge will be ruggedized and adaptable, providing for automatic synchronization to the greater cloud once communication is sufficient or reestablished. While certain DoD programs are not immediately amenable to migration to the cloud, some of these sytems may ultimately be bridged to the cloud, while others may be addressed through separate non-cloud solutions. But overall, this auto synchronization of information will ensure warfighters are retaining data, feeding it back into models, and fighting with the most recent algorithms. Doing this in a secure environment will be a force multiplier and directly support the primary goal of the cloud environment: information superiority.

Take Advantage of Resiliency in the Cloud

Enterprise cloud allows for continuity of operations and efficient failover in times of crisis and operational disruption. Cloud computing is a key component in overcoming these challenges and ensuring comprehensive mission execution, due to its distributed, scalable, and redundant nature. Executing this cloud strategy will incorporate standard approaches to leveraging cloud for this mission resiliency. The enterprise cloud will offer support for failover in times of infrastructure degradation as well as recovery from operational outages and significant cyber incidents.

The distributed, redundant nature of cloud computing overcomes another cyber challenge with its ability to failover in times of crisis. Our commercial cloud solutions will use advances in technology to automate failover, solving a major deficiency throughout the Department. DoD will only be able to ensure continuity of operations for digital services. We will accomplish this by taking advantage of multi-region and multi-availability zone (AZ) architecture, which exists natively within major cloud providers, and pairing this with the effective deployment of secure Cloud Access Points (CAPs) to cloud-based cybersecurity solutions for increased resilience. DoD cloud architectures will allow for workloads to shift from one AZ or region to another, within a single cloud provider, nearly instantaneously upon detection of the failure of a primary data center. This will be vital in the case of human-made or natural destruction of a large geographic area. The configuration of automated failover is not itself automatic. To fully achieve this capability, applications will need to be re-architected for the cloud. This will allow the Department to bypass the cost and manual effort currently required for the Department to maintain multiple instances of the same data across cloud providers or on-premises data centers, which does not provide the same level of failover as that provided by commercial cloud.

Drive IT Reform at DoD

The cloud will allow DoD to further consolidate its sprawling data center assets. The Department still has an opportunity to further rationalize and has done significant work to rationalize and reduce data centers. The cloud will provide an opportunity to accelerate and extend those consolidation opportunities, as well as the opportunity to deliver integrated Defensive Cyber Operations (DCO) and achieve efficiencies through rapid deployment of common services. An enterprise cloud perspective will enable more centralized cloud management and a broader availability of security service options for wider cloud adoption by DoD to include those DoD Components with smaller implementation staff.

Strategic Approaches and Guiding Principles

DoD requires an extensible and secure cloud environment that spans the homeland to the global tactical edge, as well as the ability to rapidly access computing and storage capacity to address warfighting challenges at the speed of relevance. Technologies such as AI and ML have the potential to fundamentally change the character of war. DoD will embrace an approach that leverages multiple cloud providers who can provide General Purpose and Fit For Purpose clouds. The interoperability of the multi-vendor and multi-cloud environment will be governed by one overarching enterprise cloud strategy. To achieve the objectives outlined above, the Department will pursue a set of guiding principles that will inform future decisions about enterprise clouds: Warfighter First, Cloud Smart-Data Smart, Leveraging Commercial Industry Best Practices, and Creating a Culture Better Suited for Modern Technology Evolution.

Warfighter First

Throughout the Department's transition to commercial cloud services, it needs to continuously test that cloud solutions are built in a manner that never puts the warfighter and his/her mission at risk. This will require the Department to rigorously red team and challenge itself with independent assessments of the cloud environment and to utilize tactical distributed computing. At all times, DoD needs to ensure that cloud is addressing the needs of improving military lethality. By constantly challenging itself around lethality with red teams, DoD can ensure that the cloud will be positioned to support the challenges of the global environment.

Cloud Smart-Data Smart

To achieve the objectives outlined above, the Department must pursue a Cloud Smart-Data Smart approach. This approach includes:

- Cloud Smart: One cloud strategy to adopt cloud solutions that streamline transformation and embrace modem capabilities for multiple clouds and missions
- Data Smart: Data transparency and visibility enabled by enterprise infrastructure, application standards, and data tagging.
- The Department seeks to leverage the decision making advantages on the battlefield enabled by AI and ML. The Department will best take advantage of these capabilities by executing this succinct,

integrated, and adaptive cloud strategy that encompasses multiple clouds and missions across the entire DoD. Systems/applications can be designed with the cloud in mind to simplify adoption and to allow for integration across the Department. Common data and application standards associated with conducting operations in the cloud, such as data normalization/tagging, transport protocols, and interfaces, will be developed to enable and encourage the adoption of enterprise solutions that navigate DoD away from custom, approaches. These standards, combined with the computing power offered by cloud, will allow the Department to function at a tempo never before seen, making informed, analytical decisions at machine speed.

Leverage Commercial Industry Best Practices

In addition to Cloud Smart-Data Smart, DoD must leverage commercial industry best practices in its approach. This includes:

- Leveraging commercial technology, capability, and innovation whenever possible
- Maximizing competition to ensure that DoD is getting the best technology and value
- Leverage industry open standards and best practices to avoid lock-in and provide maximum flexibility for future cloud advances
- Independently assessing the services delivered to ensure that the data remains secure.

The Department will leverage critical foundational technologies available in commercial cloud computing and storage, to enable innovation wherever possible, while eliminating considerable technical debt and security risk. DoD is positioning itself to get the best value in today's market of cloud computing capabilities to support warfighting and business requirements and to grow capability as industry evolves. In addition, DoD seeks to maximize competition, not only when awarding the pathfinder General Purpose cloud, but also by ensuring access to a variety of Software as a Service (SaaS) capabilities that are complementary to the General Purpose and Fit For Purpose clouds. The Department must take advantage of the advances that American private industry has made. All of this will be built into commercial pricing structures. IfDoD can adopt this commercial mindset toward cloud computing, it can incorporate commercial industry lessons learned into future architecture decisions.

Create a Culture Better Suited for Modern Technology Evolution

Finally, through this strategy, the Department seeks to create a culture that is better suited for adaptability and modem technology. This includes:

- Creating an environment where people can innovate iteratively
- Embracing enterprise solutions and navigating away from custom federated approaches
- Creating a sustainable culture and workforce that can effectively use what cloud provides
- Creating a culture that enables continuous learning from our cloud partners.

Iterative innovation is essential for successfully adapting modem technologies in an evolutionary fashion. To achieve this, DoD will embrace the use of leading modem technology quickly and more rapid prototyping of new systems. Examples include developing and deploying capabilities for DevSecOps in the cloud environment to securely develop and test software for use in the cloud and using commercial clouds to enable small and medium size companies to more effectively secure Controlled Unclassified Information (CUI). To achieve this innovation and create a culture better suited for adaptability and modem technology, the DoD workforce must change its culture. The Department must develop a cadre of technical professionals, as well as encourage technical proficiency throughout the entire Department. The Department has never built or implemented an enterprise cloud solution and therefore, recognizes the importance of finding a commercial partner to help begin the process of enterprise learning and the development of technical cloud proficiency.

IMPLEMENTATION

DoD is driving toward an enterprise cloud environment that is composed of a General Purpose cloud and multiple Fit For Purpose clouds. In addition, it should be recognized that the Department will still need non-cloud data center capability for applications that are not suited for the cloud. Over time, with the adoption of an enduring enterprise cloud strategy, the non-cloud environment should become smaller. There are two fundamental types of work that must be considered in any cloud implementation. The first is a set offundamental activities that are required to stand up a cloud platform, ready to receive applications, data, or infrastructure for cloud deployment. The second set of activities is the ongoing work to migrate existing applications or to construct new applications onto the cloud platform. Appendix A of this document is a detailed implementation plan that lays out the lines of effort that must be accomplished to fully realize the benefits of cloud computing and to effectively operate on the 21st century battlefield.

CONCLUSION

Information is a fundamental enabler for advantage on a 21st century battlefield and will enable a more lethal, resilient, and innovative Joint Force. Today, the DoD information environment is made up of multiple disjointed and stove-piped systems distributed across modern and legacy infrastructure around the globe. The data that flows through these systems is growing at an exponential rate. This has caused a litany of problems that impact warfighters,' decision makers,' and DoD staff's ability to capitalize on critical information to make timely, data-driven decisions. To address these challenges, DoD has implemented a number of cloud solutions; however, they have been built in a disjointed manner. Furthermore, DoD is starting to leverage emerging technologies, such as AI, to help manage the understanding of all the Department's data. However, the critical infrastructure that AI is being built on top of is disparate and disjointed.

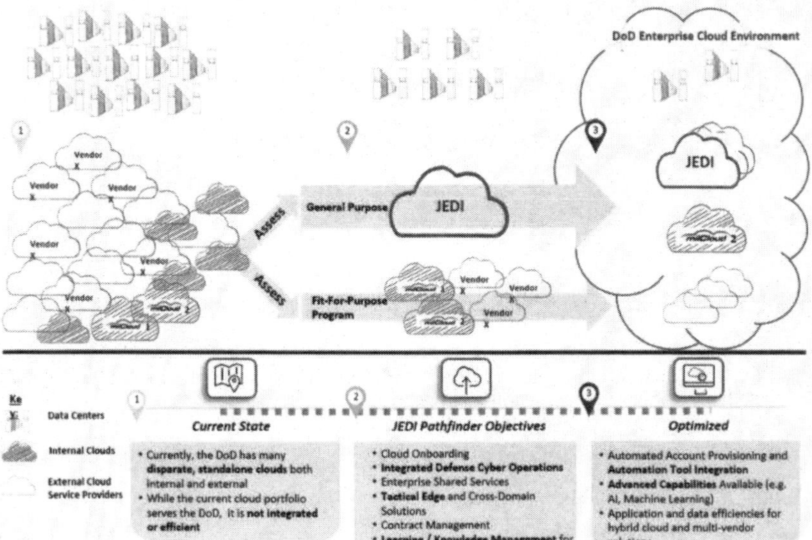

* JE DI- Joint Enterprise Defense lnfrastrucure.

Figure 1. DoD Pathfinder to Hybrid Cloud Environments and Multiple Vendors.

To overcome these challenges, DoD will utilize this guiding strategy to further develop a detailed enterprise approach for managing its data, infrastructure, and application landscape. The advent of commercial cloud has provided a powerful opportunity to address these problems. To best take advantage of the opportunity presented by commercial cloud, the Department must implement an enterprise cloud strategy. The appropriate strategy for the DoD will be to leverage a combination of General Purpose and Fit For Purpose cloud along with the advantages of multiple commercial cloud providers. To achieve the objectives outlined above, the Department will pursue a set of interrelated strategic approaches: Warfighter First, Cloud Smart-Data Smart, Leverage Commercial Industry Best Practices, and Create a Culture Better Suited for Modem Technology Evolution. DoD needs to develop an organizational construct that insures adoption of the enterprise cloud.

The time is now. DoD can no longer afford to delay its technological and cultural shift to truly modem technologies. Rapidly providing DoD access to underlying foundational technologies, like cloud computing and data storage, on a global scale is critical to national defense and in preparing DoD to fight and win wars. If the Department wants to maintain its overmatch, it will need to leverage technologies such as AI and therefore, it must urgently create an enterprise cloud environment. The strategy outlined in this document provides the approach for moving forward.

APPENDIX A: IMPLEMENTATION DETAILS

The Department of Defense is driving towards an enterprise cloud environment that is composed of a General Purpose cloud and multiple Fit For Purpose clouds. In addition, it should be recognized that the Department will still need non-cloud data center capability for applications that are not suited for the cloud. Over time, with the adoption of an enduring enterprise cloud strategy, the non-cloud environment should become smaller.

The Department's Enterprise Cloud Environment components and important considerations- migration, governance, and workforce- are described in detail below.

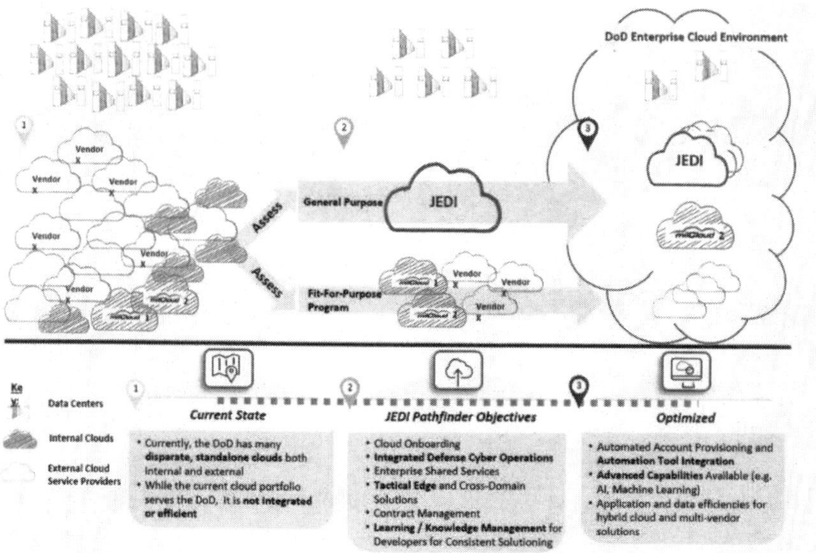

Figure A. DoD Pathfinder to Hybrid Cloud Environments and Multiple Vendors.

General Purpose Cloud: Joint Enterprise Defense Infrastructure

The Department will implement a commercial General Purpose enterprise-wide cloud solution, Joint Enterprise Defense Infrastructure (JEDI), for the majority of systems and applications. This General Purpose cloud will allow for the Department to take advantage of economies of scale, broadly provide common core services, and ensure information superiority through data aggregation and analysis.

To implement the Department's General Purpose cloud, an industry partner will be required. The complexity of this endeavor and the Department's lack of large-scale, enterprise, commercial cloud experience means that this partnership is critical to the successful standup of the enterprise General Purpose cloud.

The implementation of General Purpose cloud needs to comprehend four key tenets:

- Offer Infrastructure as a Service (IaaS) and Platform as a Service (PaaS)
- Offer separate environments at all classification levels
- Centralized computing to tactical edge computing for the warfighter
- Enable emerging technologies, such as AI

The JEDI Cloud Program will be the foundational approach to deliver the benefits of a General Purpose enterprise cloud for DoD while embracing the four tenets above. The setup of and migration to the JEDI Cloud requires the steps outlined in the Migration section. A key implementation imperative is that mission owners will be able to rapidly onboard and control their environments, but with common implementation governance supported by a contract.

Fit-for-Purpose

In situations where a General Purpose cloud solution is not capable of supporting mission needs, the Department may use a Fit For Purpose commercial solution or an on-premises cloud solution. Two examples are described here:

1. Software as a Service (SaaS): DoD software applications (e.g., email, chat, collaboration, etc.) over time will migrate to a subscription service where an indusμ-y partner will be leveraged for both applications and infrastructure.
2. The Department's milCloud 2.0 environment, a cloud-services product portfolio managed by the Defense Information Systems Agency (DISA), features an integrated suite of capabilities designed to drive agility into the development, deployment, and maintenance of secure DoD applications.

The primary implementation bias for DoD will be to utilize General Purpose cloud computing. Only when mission needs cannot be supported by General Purpose will Fit For Purpose alternatives be explored. In such a case, a mission owner will be required to submit for approval an Exception Brief to the Office of the DoD CIO describing the capability and why the General Purpose cloud service does not support their mission.

As Fit For Purpose solutions are justified, approved, and established, each Fit For Purpose cloud environment will be enacted with enterprise capabilities and scalability in mind. They should still support networking with the General Purpose cloud environment as well as with other Fit For Purpose solutions through modem commercial cloud capabilities for both inter-cloud and cross- domain communication. Secure network peering will allow for data sharing and increased visibility where required.

The Department recognizes that the commercial cloud marketplace will continue to evolve. DoD expects that cloud technology and offerings will continue to become more interoperable and seamlessly integrated, enabling lower transaction costs and better inter-cloud security features across multiple providers. DoD is best served by a robust, competitive, and innovative technology industrial base.

Cloud Migration

There are two fundamental types of work that must be considered in any cloud implementation. The first is a set of fundamental activities that are required to stand up a cloud platform, ready to receive applications, data, or infrastructure for cloud deployment. The second set of activities is the ongoing work to migrate existing applications or to construct new applications onto the cloud platform. The migration activities will need to be overseen by an ongoing governance process that assures security, application development, and data and infrastructure standards.

The fundamental activities required to stand up a cloud environment are composed of five lines of effort:

1. Technical Build - create network connectivity, encryption, data sourcing, and security services (e.g., authentication and ongoing red team engagement) on an enduring basis.
2. Governance - create stakeholder forums, policies, roadmaps, technical standards (architecture and application development), data connectivity standards, resiliency and failover standards, and migration approaches.
3. Automated Provisioning and Billing - automate the ability to quickly provision cloud resources (storage, compute, and application development) and ensure cloud costs are appropriately captured to provide financial transparency.
4. Migration Capability - create repeatable migration process, backed with qualified staff and playbooks for onboarding tenants.
5. Workforce Development - identify, train, and engage resources to create a robust and sustainable cloud workforce.

After the standup of the cloud environment and the conclusion of fundamental activities, the ongoing activities of migrating applications will take over. The magnitude of effort required to stand up a General Purpose cloud at the scale and complexity of the Department is initially best served through a single provider that will allow DoD to maximize pace and minimize risk. The effort required to migrate applications will vary greatly from system to system. Migrating to a cloud environment is not typically as simple as "lift and shift." The migration process will be defined in the DoD's Cloud Migration Playbook and will include the many different paths to realize cloud. The Department will closely monitor the initial efforts to migrate into the Department's General Purpose enterprise cloud. The lessons learned from the various migrations will inform the regular refinement of the playbook, which will provide a consistent and repeatable process for mission owners to apply to their respective systems and applications.

Organizations within DoD that have previously implemented their own cloud will work with the Office of the DoD CIO to determine the best way to integrate their efforts with the Department's enterprise cloud strategy. Where it makes sense that a standalone cloud environment should be migrated to the Department's General Purpose cloud, a thoughtful migration approach will be developed that does not disrupt existing contracts. If it is determined that an existing standalone cloud should be retained, it will be given a formal Fit For Purpose designation and will be expected to adhere to the DoD's enterprise cloud policies.

Governance and Organization

Oversight and governance for the initial build out of this enterprise cloud strategy will be led by the Office of the DoD CIO. At some future date, once the General Purpose cloud environment is fully implemented and Fit For Purpose implementations have matured, it is possible that overall leadership could be transitioned to a different organization inside DoD. The DoD CIO will establish an enterprise cloud organization with appropriate leadership and the required governance forums to ensure that overall objectives and implementation plans as described in this strategy are enacted. The DoD CIO will leverage existing governance forums to the greatest extent possible.

The DoD CIO will organize forums that bring together all lessons learned and find ways to integrate into DoD policies, procedures, and acquisition strategies moving forward. These forums will allow the Department to do regular checks on cloud adoption progress and course correct quickly. The forums will develop detailed implementation plans on rationalization, assessment, planning, and budgeting for accelerating the digital environment to migrate into the enterprise cloud solution.

These forums must be the catalyst to aggressively move the Department to embrace the use of cloud. The DoD CIO in conjunction with the governance forums will provide written guidance on the process and what factors are key to assessing a system for General Purpose or Fit For Purpose hosting. The Department must strive for a Cloud First bias on all future application development/implementations. Organizations should move toward modernization by implementing "cloud native" applications, meaning that the architecture of the system can make use of the scalable, dynamically elastic, distributed nature of cloud computing platforms. Leaving systems running on legacy infrastructure or using other legacy technology must be the exception, not the standard.

Additionally, these forums will work through the Office of the Under Secretary of Defense for Acquisition and Sustainment (A&S) to address current regulations that govern acquisitions to fully take advantage of modem utility/consumption-based services and to enhance contracting capabilities and ATO processes to enable reuse of PaaS/SaaS and cloud-based applications. Working through Cost Assessment and Program Evaluation (CAPE), the forums will provide early insight into annual Planning, Programming, Budgeting, and Execution (PPBE) for cloud activity in the Department. Additionally, these forums will work with the Office of Chief Management Officer (CMO) for application of relevant data standards and governance processes in cloud activities.

As JAIC matures, a key organizational imperative will be to ensure that the requirements of JAIC and enterprise cloud are being jointly integrated.

Workforce Considerations

In today's world, our adversaries are working to develop new capabilities that leverage the advantages of cloud. Therefore, we must ensure, that as we migrate and become more dependent on the cloud, that we are organically growing our cloud technical skills to outpace our adversaries. We can never lose sight that cloud is a key enabler for emerging technologies, such as AL The future DoD cloud workforce must grow organic technical capabilities by building a more diverse and balanced workforce among military and civilian components. DoD's workforce must obtain a basic level of cloud proficiency in order to most effectively exploit the benefits of cloud. Just as every Marine is a rifleman, every DoD employee must have basic cloud awareness to effectively operate on the 21st century battlefield.

In: New Strategies in Cloud Computing ISBN: 978-1-53616-907-2
Editor: Luc Larivière

Chapter 4

THE DOD'S JEDI CLOUD PROGRAM*

Heidi M. Peters

In September 2017, Deputy Secretary of Defense (DSD) Patrick Shanahan issued a memorandum calling for the accelerated adoption of a Department of Defense (DOD)-wide cloud computing system. Under the Joint Enterprise Defense Infrastructure (JEDI) Cloud program, DOD seeks to "acquire a ... cloud services solution that can support Unclassified, Secret, and Top Secret requirements," with a focus on commercially available services. Significant industry and congressional attention has been focused on the JEDI Cloud contract.

WHAT IS CLOUD COMPUTING?

Broadly speaking, *cloud computing* refers to the practice of remotely storing and accessing information and software programs through the internet, instead of storing data on a computer's hard drive or accessing it

* This is an edited, reformatted and augmented version of Congressional Research Service, Publication No. IN10990, Version 6, dated December 12, 2018.

through an organization's intranet. It relies on *cloud infrastructure*, a collection of hardware and software that may include components such as servers and a network. This infrastructure can be deployed *privately* to a select user group, or *publicly* through commercial services available to the general public.

WHAT IS THE CURRENT STATUS OF DOD'S ADOPTION OF CLOUD SERVICES?

DOD maintains more than 500 public and private cloud infrastructures that support Unclassified and Secret requirements. DOD has described its current cloud services use as "decentralized," creating "additional layers of complexity for managing data and services at an enterprise level." In a statement accompanying the release of the JEDI Cloud Request for Proposal (RFP) on July 26, 2018, DOD Chief Information Officer (CIO) Dana Deasy noted that the department requires an enterprise-wide cloud "that allows for data-driven decision making [and] enables DOD to take advantage of our applications and data resources" to provide worldwide support for DOD operations.

WHAT POLICIES APPLY TO DOD CLOUD ACQUISITIONS?

While the Federal Acquisition Regulation (FAR) does not specifically provide acquisition guidance for cloud computing services, select sections (such as FAR Part 39, Acquisition of Information Technology) may apply. Other government-wide policies for cloud products and services, such as the Federal Risk and Authorization Management Program (FedRAMP), may also apply.

DOD policies for acquiring cloud services are prescribed by Defense Federal Acquisition Regulation Supplement Subpart 239.76, which states that DOD must generally acquire cloud services using commercial terms and conditions consistent with federal law and DOD's needs. A contract to acquire cloud services may generally only be awarded to a provider (e.g., a prime contractor or subcontractor) with provisional Defense Information Security Agency (DISA) authorization to provide such services, consistent with the current version of the DOD Cloud Computing Security Requirements Guide. DOD Instruction 5000.74, *Defense Acquisition of Services*, specifies that all cloud services must have an Authority to Operate (see also DOD Instruction 8510.01, *Risk Management Framework for DOD Information Technology*).

DOD Memorandum *Updated Guidance on the Acquisition and Use of Commercial Cloud Computing Services*, issued on December 15, 2014, provides additional guidance for the acquisition of commercial cloud services.

HOW IS THE JEDI CLOUD CONTRACT STRUCTURED?

Through the JEDI Cloud contract, DOD is conducting a full and open competition, which will result in a single award Indefinite Delivery/Indefinite Quantity (ID/IQ) firm-fixed price contract for commercial items (i.e., cloud infrastructure and computing services).

What is an ID/IQ Contract?

An *ID/IQ contract* allows the U.S. government to obtain an unspecified quantity of supplies or services over an unspecified period of time. An indefinite-quantity contract can also be referred to as a *task-order contract*. A *task-order contract* does not procure a firm quantity of services (other than a minimum or maximum quantity) and allows the issuance of orders for the performance of tasks (i.e., *task orders)* under the contract.

The period of performance is structured as a 2-year base ordering period, with three additional option periods (two 3-year options and one 2-year option), for a potential total of 10 years.

Anticipated JEDI Cloud Contract Period of Performance

Period of Performance	**Timeframe**
Base ordering period (2 years, guaranteed)	April 17, 2019-April 16, 2021
Option #1 (3 years, if exercised)	*April 17, 2021-April 16, 2024*
Option #2 (3 years, if exercised)	*April 17, 2024-April 16, 2027*
Option #3 (2 years, if exercised)	*April 17, 2027-April 16, 2029*

Source: JEDI Cloud RFP, "Combined Synopsis/Solicitation for Commercial Items," updated August 31, 2018.

DOD wants the JEDI Cloud to provide services comparable to those made available through commercial cloud services and is requiring any successful offeror to provide rapid deployment of new commercially available cloud-related services to JEDI Cloud users. DOD also expects ongoing parity with public commercial prices.

DOD indicated that the minimum guaranteed award is $1 million. The contract has a maximum ceiling of $10 billion across the potential 10-year period of performance. Under an ID/IQ contract, the government is only required to purchase the minimum amount specified in the contract, and may ultimately choose not to reach the contract ceiling.

The RFP closed on October 9, 2018; DOD is currently reviewing received bids.

How Has Industry Reacted?

DOD received more than 1,500 comments in response to multiple draft RFPs; companies including Amazon, Google, IBM, and Microsoft expressed interest in competing. However, DOD's proposed acquisition strategy also sparked resistance from those who opposed DOD's intent to award the contract to a single company. This concern led some industry associations to publicly contest a single award, arguing that it would be inconsistent with broader federal cloud computing strategy, and could unfairly restrict future competition for DOD cloud services. For example, the trade group ITAPS (IT Alliance for Public Sector) sent a letter to the House and Senate Armed Services committees stating in part that the

> ... deployment of a single cloud conflicts with established best practices and industry trends in the commercial marketplace, as well as current law and regulation, which calls for the award of multiple task or delivery order contracts.... Further, the speed of adoption of innovative commercial solutions, like cloud, is facilitated by the use of these best practices.

On October 8, 2018, Google announced that it would not be submitting a bid for the contract, citing possible conflict with its corporate principles, along with DOD's plans to award the contract to a single vendor, among its reasons for withdrawing.

Oracle America and IBM both filed pre-award bid protests against the JEDI Cloud solicitation; GAO denied Oracle America's protests on November 14, 2018, and denied IBM's protests on December 11, 2018. Subsequent to GAO's dismissal of its protests, Oracle America filed a bid protest lawsuit with the U.S. Court of Federal Claims (Oracle America Inc. v. U.S., case number 1:18-cv-01880).

How Has DOD Responded to Industry Concerns?

DSD Shanahan and other DOD officials have reportedly described JEDI Cloud as a "pathfinder" intended to provide a model for DOD's future transition of legacy IT systems to the cloud. In its May 2018 report to Congress DOD indicated that the JEDI Cloud contract would include

> ... multiple mechanisms to ... maximize DOD's flexibilities going forward.... Option periods ... will only be exercised if doing so is the most advantageous method for fulfilling the DOD's requirements when considering the market conditions at the time of option exercise.

DOD noted that a multiple-award contract would require competition for each task order issued under the JEDI Cloud ID/IQ, and would therefore be subject to standard timeframes associated with the DOD acquisition process, which could "prevent DOD from rapidly delivering new capabilities and improved effectiveness ... that enterprise-level cloud computing can enable."

DOD has also said that it "expects to maintain contracts with numerous cloud providers to access specialized capabilities not available under the JEDI Cloud contract."

WHAT ACTIONS HAS CONGRESS TAKEN?

Section 8137 of P.L. 115-245, which provided FY2019 DOD appropriations, prevents the obligation or expenditure of FY2019 funds to "migrate data and applications to the proposed [JEDI] ... cloud computing services" until 90 days after the Secretary of Defense submits to the congressional defense committees (1) a plan to establish a DOD-wide budget accounting system for funds requested and expended for cloud services, as well as funds requested and expended to migrate to a cloud environment; and (2) a detailed description of DOD's strategy to implement enterprise-wide cloud computing.

On October 22, 2018, two members of the House Appropriations Committee asked the DOD Inspector General (IG) to investigate the development of requirements and the RFP process for the JEDI Cloud program; the DOD IG is reviewing the request.

In: New Strategies in Cloud Computing
Editor: Luc Larivière
ISBN: 978-1-53616-907-2

Chapter 5

CROSS-BORDER DATA SHARING UNDER THE CLOUD ACT*

Stephen P. Mulligan

SUMMARY

Law enforcement officials in the United States and abroad increasingly seek access to electronic communications, such as emails and social media posts, stored on servers and in data centers in foreign countries. Because the architecture of the internet allows technology companies to store data at a great distance from the physical location of their customers, electronic communications that could serve as evidence of a crime often are not housed in the same country where the crime occurred. This disconnect has caused governments around the world, including the United States, to seek data stored outside their territorial jurisdictions. In the Clarifying Lawful Overseas Use of Data (CLOUD) Act, Congress enacted one of the first major changes in years to U.S. law governing cross-border access to electronic communications held by private companies.

* This is an edited, reformatted and augmented version of a Congressional Research Service publication R45173, prepared for Members and Committees of Congress dated April 23, 2018.

The CLOUD Act has two major components. The first facet addresses the U.S. government's ability to compel technology companies to disclose the contents of electronic communications stored on the companies' servers and data centers overseas. The Stored Communications Act (SCA) mandates that certain technology companies disclose the contents of electronic communications pursuant to warrants issued by U.S. courts based on probable cause that the communications contain evidence of a crime. But a dispute arose over whether warrants issued under the SCA could compel disclosure of data held outside the territorial jurisdiction of the United States. While the Supreme Court was set to resolve this issue in *United States v. Microsoft*, the CLOUD Act amended the SCA to require that technology companies provide data in their possession, custody, or control in response to an SCA warrant—regardless of whether the data is located in the United States. On April 17, 2018, the Supreme Court ruled that the change in law mooted the *Microsoft* case.

The second facet of the CLOUD Act addresses the reciprocal issue of foreign governments' ability to access data in the United States as part of their investigation and prosecution of crimes. Prior to the CLOUD Act, foreign nations seeking data in the United States were required to request the assistance of the U.S. government through either mutual legal assistance treaties (MLATs) or judicial instruments known as letters rogatory. Requests under either instrument are reviewed by U.S. courts before disclosure to the foreign nation can be authorized, but U.S. and foreign officials criticized the processes as inefficient and unable to accommodate the increasing number of data requests in the digital era.

The CLOUD Act responds to calls for modernization by authorizing the executive branch to conclude a new form of international agreement through which select foreign governments can seek data directly from U.S. technology companies without individualized review by the U.S. government. Agreements authorized by the CLOUD Act would remove legal restrictions on certain foreign nations' ability to seek data directly from U.S. providers in cases involving "serious crimes" when not targeting U.S. persons, provided the Executive has determined that the foreign nation's laws adequately protect privacy and civil liberties, among other requirements. While the CLOUD Act conditions approval of covered agreements upon a host of restrictions, commentators debate whether these agreements will provide adequate protections for privacy, human rights, and civil liberties.

Law enforcement officials in the United States and abroad increasingly seek access to electronic communications, such as emails and social media posts, stored on servers and in data centers located in foreign countries.[1] The architecture of the internet allows technology companies significant flexibility as to the geographic location where they may store collected data.[2] As a result, electronic communications that may be evidence of a crime are not necessarily housed in the same country where the crime occurred.[3] This disconnect has caused governments around the world, including the United States, to seek data stored outside their territorial jurisdictions in the course of law enforcement investigations.[4] It also has led to debate over the extent to which national governments can compel private companies to disclose data stored in foreign nations and the degree to which civil liberties and privacy concerns should inform the proper procedure for sharing such data.[5] In the United States, this debate largely has centered on the Stored

---

[1] *See, e.g.*, Andrew Keane Woods, *Against Data Exceptionalism*, 68 Stan. L. Rev. 729, 742-45 (2016) (analyzing trends of increased government demands for data located outside a nation's territorial jurisdiction); *Data Stored Abroad: Ensuring Lawful Access and Privacy Protection in the Digital Era: Hearing Before the H. Comm. on the Judiciary*, 115th Cong. 1 (2017) [hereinafter *Data Stored Abroad Hearing*] (statement of Richard W. Downing, Acting Deputy Assistant Att'y Gen., U.S. Dep't of Justice), https://judiciary.house.gov/wp-content/uploads/2017/06/Downing- Testimony.pdf [hereinafter Downing Statement] (outlining challenges to U.S. and foreign government efforts to obtain data overseas).

[2] *See, e.g.*, Riley v. California, 134 S. Ct. 2473, 2490-91 (2014) ("Cloud computing is the capacity of Internet-connected devices to display data stored on remove servers rather than on the device itself.");Woods, *supra* note 1, at 739 ("[O]ne of the greatest societal and technological shifts *In re*cent years has been the move from storing data on a local machine—such as a cell phone or computer—to storing that data remotely on faraway servers, which can be accessed by a network such as the Internet.").

[3] *See, e.g., Data Stored Abroad Hearing*, *supra* note 1 (statement of Paddy McGuinness, Deputy Nat'l Sec. Advisor, U.K.), https://judiciary.house.gov/wp-content/uploads/2017/06/ McGuinness-Testimony.pdf [hereinafter McGuinness Statement] (discussing the need for U.K. law enforcement access to data stored in the United States); *Hearing on International Conflicts of Law Concerning Cross Border Data Flow and Law Enforcement Requests Before the H. Comm. on the Judiciary*, 114th Cong. 22, 57-59 (2016) [hereinafter *International Conflicts of Law Hearing*] (statement of Brad Smith, President and Chief Legal Officer, Microsoft Corp.) [hereinafter Smith Statement] (discussing French requests for data stored by Microsoft following a 2015 terrorist attack in Paris).

[4] *See supra* notes 1-3. *See also infra* § "*United States v. Microsoft Corp.* and the CLOUD Act" (discussing the United States efforts to obtain data in Ireland); *International Conflicts of Law Hearing*, *supra* note 3, at 17-18 (statement of David Bitkower, Principal Assistant Deputy Att'y Gen., U.S. Dep't of Justice) [hereinafter Bitkower Statement] (listing examples of evidence gathered from American technology companies that was critical to solving crimes overseas); Peter Swire et al., *A Mutual Legal Assistance Case Study: The United States and France*, 34 Wis. Int'l L.J. 323, 327 (2016) (discussing "how the globalization of data is affecting even routine criminal investigations").

[5] *Compare, e.g.*, Jennifer Daskal, *The Un-Territoriality of Data*, 125 Yale L.J. 326, 329 (2015) (contending that the unique nature of data and the "physical disconnect between the location of data and the location of its user" undermines traditional notions of territorial sovereignty), *with* Woods, *supra* note 1, at 756-63 (arguing that data is compatible with existing conceptions of sovereignty and jurisdiction). *See also infra* § "Commentary on the

Communications Act (SCA),[6] which is part of the broader Electronic Communications Privacy Act (ECPA).[7] Although the SCA generally prohibits certain technology companies from disclosing the contents of electronic communications to third parties,[8] it mandates disclosure to the U.S. government pursuant to a warrant based on probable cause that the communications contain evidence of a crime.[9] In *United States v. Microsoft Corp.*, the Supreme Court was set to address whether the United States could compel Microsoft to release emails housed in a data center in Ireland through a warrant issued under the SCA.[10] But less than one month after oral argument, Congress passed and the President signed into law the Clarifying Lawful Overseas Use of Data Act (CLOUD Act) as part of the Consolidated Appropriations Act, 2018.[11] The CLOUD Act amends the SCA and requires service providers subject to the SCA[12] to release data in their possession, custody, or control in response to an SCA warrant—regardless of whether the data is located in the United States.[13] After the U.S. government obtained a new warrant for the emails held in Ireland under the authority of the CLOUD Act, the Supreme Court deemed *Microsoft* moot.[14]

A second facet of the CLOUD Act addresses the reciprocal issue of foreign governments' desire to access data in the United States as part of their investigation and prosecution of crimes.[15] Prior to the CLOUD Act, foreign nations seeking data in the United States generally were required to request the assistance of the U.S. government through either procedures established by mutual legal assistance treaties (MLATs) or judicial requests known as letters rogatory.[16] Requests under either

---

CLOUD Act" (discussing commentary regarding the extent to which cross-border data sharing regimes should provide safeguards for privacy, human rights, and civil liberties).

6 *See* 18 U.S.C. §§ 2701-2712.

7 *See* P.L. 99-508, 100 Stat. 1848 (1986).

8 *See* 18 U.S.C. § 2702(a).

9 *Id.* § 2703(a).

10 *See* No. 17-2, 548 U.S. __, 2018 WL 1800369, slip op. at 2 (U.S. Apr. 17, 2018) (per curiam).

11 *See* Consolidated Appropriations Act, 2018, P.L. 115-141, div. V [hereinafter CLOUD Act].

12 As discussed in more detail below, the SCA applies to a provider of an "electronic communications service," defined in 18 U.S.C. § 2510(15), and a "remote computing service," defined in 18 U.S.C. § 2711(2). *See infra* "Overview of ECPA and the SCA." Unless otherwise indicated, the terms "service providers" or "providers" in this chapter reference both entities covered by the SCA.

13 CLOUD Act § 103 (adding 18 U.S.C. § 2713).

14 *See* No. 17-2, 548 U.S. __, 2018 WL 1800369, slip op. at 2 (U.S. Apr. 17, 2018) (per curiam) (vacating and remanding with instructions to dismiss as moot).

15 *See* CLOUD Act § 102(3) (discussing foreign governments' need to "access electronic data held by communications- service providers in the United States" in the congressional findings). *See also infra* § "Executive Agreements Authorized by the CLOUD Act."

16 *See* T. Markus Funk, Mutual Legal Assistance Treaties and Letters Rogatory: A Guide for Judges 1 (Fed. J. Center 2014), https://www.fjc.gov/sites/default/files/2017/ MLAT-LR-

instrument are reviewed by U.S. courts before disclosure to the foreign nation is authorized, but U.S. and foreign officials have criticized these processes as inefficient and unable to accommodate the increasing cross-border data demands in the digital era.[17]

The CLOUD Act responds to calls for modernization by authorizing the executive branch to conclude a new form of international agreement[18] through which select foreign governments can seek data directly from U.S. technology companies without undergoing individualized review by the U.S. government.[19] Agreements authorized by the CLOUD Act would remove legal restrictions on certain foreign nations' ability to seek data directly from U.S. providers in cases involving "serious crimes" when not targeting U.S. persons, provided that the United States has determined that the foreign nation's laws adequately protect privacy and civil liberties, among other requirements.[20]

This chapter reviews the development of cross-border data sharing laws in criminal matters in the United States.[21] It begins with an overview of ECPA and the SCA.[22] Next, the report discusses the questions raised in the *Microsoft* litigation and the impact of the CLOUD Act on those

---

Guide-Funk-FJC-2014.pdf; Woods, *supra* note 1, at 748-49. While MLATs and letters rogatory have been the standard legal avenues for seeking cross-border data, some information can be provided through informal channels, such as cooperative exchange between investigators. *See* FUNK, *supra*, at 23.

[17] *See, e.g.*, President's Review Grp. on Intelligence & Commc'ns Techs., Liberty and Security in a Changing World: Report and Recommendations 227 (2013) [hereinafter President's Review Group] ("The MLAT process . . . is too slow and cumbersome."); Downing Statement, *supra* note 1, at 7 ("[T]he [mutual legal assistance] process can lack the requisite efficiency for time-sensitive investigations and other emergencies, making it an impractical alternative to SCA warrants in many cases."); McGuinness Statement, *supra* note 3 ("It is widely acknowledged that MLAT processes are too slow for rapidly developing counter terrorism and serious crime investigations.").

[18] As used in this chapter, the term "international agreement" is intended to be a blanket term that includes all agreements between the United States and foreign nations that are intended to be binding under international law. *Accord* Restatement (Fourth) of Foreign Relations Law: Treaties, Tentative Draft No. 2, § 102 cmt. a (2017).

[19] *See infra* § "Executive Agreements Authorized by the CLOUD Act."

[20] *See id.*

[21] Because this chapter focuses on data sharing in the context of criminal investigations, it does not address other, unrelated forms of information sharing, such as information sharing within an industry or with the government following a cyberattack, see CRS In Focus IF10163, *Cybersecurity and Information Sharing*, by N. Eric Weiss, or information shared among private companies for commercial purposes, see *Facebook, Social Media Privacy, and the Use and Abuse of Data, Hearing Before the S. Comm. on Commerce, Science, and Transportation* 115th Cong. (Apr. 10, 2018).

[22] *See infra* § "Overview of ECPA and the SCA." Although constitutional provisions such as the Fourth Amendment are relevant to government access to personal data as part of a criminal investigation, *see* United States v. Warshak, 631 F.3d 266 (6th Cir. 2010) (holding that the government must obtain a warrant to access certain stored emails), the focus of this chapter is on statutory protections.

issues.[23] Finally, the report examines the new form of international agreements authorized by the CLOUD Act and the commentary on the benefits and drawbacks of the potential new international data sharing agreements.[24]

OVERVIEW OF ECPA AND THE SCA

Enacted in 1986, ECPA is one of the primary federal laws regulating disclosure of electronic communications held by private entities.[25] ECPA is structured on three main titles. Title I, commonly referred to as the Wiretap Act, governs the interception of real-time wire, oral, or electronic communications.[26] Title II added a new chapter to the *United States Code* entitled "Stored Wire and Electronic Communications and Transactional Records Access," and generally is referred to as the Stored Communications Act or SCA.[27] The SCA applies to many forms of electronic communications and associated data, including emails;[28] text messages;[29] private messages, wall postings, and other comments made on or via social media sites;[30] and private YouTube videos.[31] Title III of ECPA regulates the use of a pen register, a device that allows users to capture the routing information associated with communications, such as telephone numbers dialed.[32] Each title in ECPA contains restrictions on the circumstances in which the relevant data can be used or disclosed.[33]

---

[23] *See infra* § "*United States v. Microsoft Corp.* and the CLOUD Act."
[24] *See infra* § "Executive Agreements Authorized by the CLOUD Act."
[25] *See* P.L. 99-508, 100 Stat. 1848 (1986).
[26] *See id.* tit. I, 100 Stat. at 1848-59 (codified in 18 U.S.C. §§ 2510-2521).
[27] *Id.* at 1860.
[28] *See* Theofel v. Farey-Jones, 359 F.3d 1066, 1077 (9th Cir. 2004), *cert denied* 543 U.S. 813 (2004).
[29] *See* Quon v. Arch Wireless Operating Co., Inc., 529 F.3d 892, 901 (9th Cir. 2008), *rev'd on Fourth Amendment grounds sub nom.* Quon v. City of Ontario, 560 U.S. 746 (2010).
[30] *See* Crispin v. Christian Audigier, 717 F. Supp. 2d 965, 980, 989 (C.D. Cal. 2010).
[31] *See* Viacom Intern. Inc. v. YouTube Inc., 253 F.R.D. 256, 264 (S.D.N.Y. 2008).
[32] P.L. 99-508, tit. III, 100 Stat. 1848, 1868-73 (codified in 18 U.S.C. §§ 3121-3127).
[33] *See* 18 U.S.C. §§ 2511(1), 2702; 3121. For additional analysis of ECPA and its provisions, see CRS Report R41733, *Privacy: An Overview of the Electronic Communications Privacy Act*, by Charles Doyle, and CRS Report R41734, *Privacy: An Abridged Overview of the Electronic Communications Privacy Act*, by Charles Doyle.

As technology has evolved since ECPA's enactment in 1986, law enforcement has shifted its primary focus from the interception of live communications pursuant to the Wiretap Act to seeking the now-common forms of stored communications governed by the SCA.[34] But the SCA does not apply the same provisions to every communication or data that falls under its ambit. Rather, the scope of the SCA may be impacted by whether the law is applied to a provider of "electronic communication services" (ECS) or "remote computing services" (RCS).[35] Although some SCA requirements vary depending on the provider,[36] the act has two core components that apply to both forms of provider: (1) prohibitions on disclosure of certain data and (2) mandatory disclosure provisions.[37]

Prohibitions on Disclosure under the SCA

The first facet of the SCA is a restriction on providers' ability to share customers' electronic communications and their related records and information. Restrictions differ depending on the data at issue.[38] For the *contents* of electronic communications (e.g., the body of an email), the SCA prohibits disclosure to "any person or entity," absent an exception, provided certain technical requirements are met.[39] The SCA also prohibits

---

[34] *See* Orin Kerr, *The Next Generation Communications Privacy Act*, 162 U. Pa. L. Rev. 373, 394 (2014).

[35] *See* 18 U.S.C. § 2702(a)(1)-(2).

[36] A provider of ECS allows its customers "to send or receive wire or electronic communications." *Id.* § 2510(15). A provider of RCS provides "computer storage or processing services by means of an electronic communication system." *Id.* § 2711(2).

[37] *See infra* §§ "Prohibitions on Disclosure Under the SCA"; "Mandatory Disclosure Under the SCA."

[38] *See* 18 U.S.C. § 2702.

[39] Providers of ECS may not disclose the contents of communication "while in electronic storage." *Id.* § 2702(a)(1). Providers of RCS may not disclose the contents of a communication that is "carried or maintained" by the service, provided two additional conditions are satisfied. *Id.* § 2702(a)(2). First, the communication must be maintained "on behalf of, and received by means of electronic transmission from (or created by means of computer processing of communications received by means of electronic transmission from), a subscriber or customer of such service." *Id.* § 2702(a)(2)(A). Second, the communication must be maintained "solely for the purpose of providing storage or computer processing services to such subscriber or customer, if the provider is not

both categories of provider from disclosing a "record or other information pertaining to a subscriber to or customer of such service" to the U.S. government.[40] This prohibition, which concerns non-content information or "metadata," does not prohibit disclosure to private entities or foreign governments.[41] The SCA enumerates several exceptions to the prohibition on disclosure of both content[42] and non-content communications.[43]

Mandatory Disclosure under the SCA

The second major component of the SCA is its rules that *require* providers to disclose customer communications and related records to the U.S. government.[44] The SCA establishes a tiered system with differing procedures and standards governing when the U.S. government can

---

authorized to access the contents of any such communications for purposes of providing any services other than storage or computer processing." *Id.* § 2702(a)(2)(B).

[40] *Id.* § 2702(a)(3) ("a provider of remote computing service or electronic communication service to the public shall not knowingly divulge a record or other information pertaining to a subscriber to or customer of such service (not including the contents of communications covered by paragraph (1) or (2)) to any governmental entity."). The SCA defines "government entity" as "a department or agency of the United States or any State or political subdivision thereof." *Id.* § 2711(4).

[41] *Id.* § 2702(c)(6). Other federal or state laws may prohibit disclosure of particular classes of non-content information to foreign governments or private entities even if the SCA does not. *See, e.g., id.* § 2710 (restricting disclosure of "prerecorded video cassette tapes or similar audio visual materials"); 20 U.S.C. § 1232g(b) (restricting the disclosure of "education records" by education agencies or institutions that receive federal funds).

[42] Among other exceptions enumerated in 18 U.S.C. § 2702(b), providers may divulge the content of communications: to an addressee or intended recipient; as may be necessary incident to the rendition of the service or the protection of the rights of property of the provider of that service; or to the U.S. government, if the provider, in good faith, believes that an emergency involving danger of death or serious physical injury to any person requires disclosure without delay.

[43] Exceptions to the prohibition on disclosure of non-content data are listed in 18 U.S.C. § 2702(c). These exceptions include, among things, disclosure (1) with the lawful consent of the customer or subscriber; (2) as may be necessarily incident to the rendition of the service or to the protection of the rights or property of the provider of that service; (3) to the U.S. government, if the provider, in good faith, believes that an emergency involving danger of death or serious physical injury to any person requires disclosure without delay; (4) to the National Center for Missing and Exploited Children; and (5) to any non-U.S.-government person or entity.

[44] *See* 18 U.S.C. § 2703.

demand that providers divulge stored communications.[45] As described below, the SCA's standards for mandatory disclosure depend on a number of factors, including, among other things, the type of data sought; whether an ECS or RCS holds the data; the length of time the data has been stored; whether the data is content or non-content; and whether advanced notice has been given to the customer.[46] The multitude of relevant factors can make the determination of whether disclosure is mandatory a complex and fact-specific evaluation.[47]

At the highest level, the SCA requires the U.S. government to obtain a warrant if the government seeks access from an ECS provider to the *content* of a communication that has been in "electronic storage" for 180 days or less.[48] A warrant may be issued only if the U.S. government demonstrates probable cause that the communications sought establish evidence of a crime.[49] If the communication has been stored for longer than 180 days, or if it is being "held or maintained" by an RCS "solely for the

---

[45] *See infra* notes 48-53.

[46] *See id.*

[47] For example, whether disclosure of email content is required may depend on, among other factors, the technical architecture of the email system and whether the intended recipient opened the email. *See* United States v. Weaver, 636 F. Supp. 2d 769, 771 (C.D. Ill. 2009) (discussing how the SCA's mandatory disclosure requirements differ when applied to a "web-based email system" as compared to other email systems); Orin K. Kerr, *A User's Guide to the Stored Communications Act, and a Legislator's Guide to Amending It*, 72 Geo. Wash. L. Rev. 1208, 1220-24 (2004) (providing background on ECPA). (discussing the application of the SCA's mandatory disclosure provisions to various forms of email in transit and in storage).

[48] 18 U.S.C. § 2703(a). "Electronic storage" is defined as "(A) any temporary, intermediate storage of a wire or electronic communication incidental to the electronic transmission thereof; and (B) any storage of such communication by an electronic communication service for purposes of backup protection of such communication." 18 U.S.C. § 2510(17). The case law generally holds that a user-opened email stored solely on the email provider's server is not in "electronic storage." *See* Theofel v. Farey-Jones, 359 F.3d 1066, 1077 (9th Cir. 2004) ("A remote computing service might be the only place a user stores his messages; in that case, the messages are not stored for backup purposes."); Fraser v. Nationwide Mut. Ins. Co., 135 F. Supp. 2d 623, 636 (E.D. Penn. 2001) ("[M]essages that are in post- transmission storage, after transmission is complete, are not covered by part (B) of the definition of 'electronic storage'").

[49] *See* 18 U.S.C. § 2703(a) (requiring that any warrant issued under the SCA be "issued using the procedures described in the Federal Rules of Criminal Procedure (or, in the case of a State court, issued using State warrant procedures) by a court of competent jurisdiction"); Fed. R. Crim. P. 41(d)(1) ("[A] magistrate judge—or if authorized . . . a judge of a state court of record—must issue the warrant if there is probable cause to search for and seize a person or property or to install and use a tracking device.").

purpose of providing storage or computer processing services," the government can use a subpoena or a court order under 18 U.S.C. § 2703(d), provided notice is given to the customer.[50] To obtain an order under this section—known as a Section 2703(d) order—the applicant must prove "specific and articulable facts, showing that there are reasonable grounds to believe that the contents of a[n] . . . electronic communication . . . are relevant and material to an ongoing criminal investigation."[51]

In addition to the content of communications, the SCA permits access to non-content information with a warrant, but the government also may use a subpoena or a Section 2703(d) order to provide the customer notice.[52] To access basic subscriber information, including the customer's name, address, phone number, length of service, and means of payment (including bank account numbers), the government may follow the more stringent requirements for obtaining a warrant or a Section 2703(d) order, but it also can use an administrative subpoena, which requires no prior authorization by a judicial officer or notice to the customer.[53]

United States v. Microsoft Corp. and the CLOUD Act

While the complexities of the SCA coupled with major changes in technology have led some to call for broad reforms to the law,[54] one discrete issue—the extraterritorial application of the SCA—became the subject of particular interest as a result of a 2016 federal appellate court decision.[55] As noted above, the SCA mandates that service providers

---

[50] *See* 18 U.S.C. § 2703(a); § 2703(b)(1)(B).

[51] *Id.* § 2703(d).

[52] *See id.* § 2703(c).

[53] *See id.*

[54] *See, e.g.*, Kerr, *supra* note 34, at 376-78; Caroline Lynch, *ECPA Reform 2.0. Previewing the Debate in the 115th Congress*, Lawfare (Jan. 30, 2017), https://www.lawfareblog.com/ecpa-reform-20-previewing-debate-115th-congress.

[55] *See* Matter of Warrant to Search a Certain E-Mail Account Controlled and Maintained by Microsoft Corporation, 829 F.3d 197, 222 (2d Cir. 2016) [hereinafter Matter of Warrant],

disclose the content of electronic communications when the government obtains a warrant based on probable cause.[56] In 2013, federal law enforcement officials sought an SCA warrant requiring Microsoft to disclose all emails and other information associated with an account with one of its customers.[57] After finding that the United States demonstrated probable cause that the account was being used to further illegal drug trafficking, a United States magistrate judge issued a warrant requiring Microsoft to disclose the contents of an email account and all records or information associated with the account "[t]o the extent that the information . . . is within [Microsoft's] possession, custody, or control."[58]

Microsoft complied with the portion of the warrant seeking metadata about the user's account (e.g., the name, IP address, and telephone number associated with the account), which was stored in the United States, but it determined that the contents of the user's emails were held in a data center in Dublin, Ireland.[59] Microsoft stores its users' emails in one of its many data centers around the world—most often the one closest to where users state they are from when signing up for the email service.[60] Although Microsoft did not dispute that it had the ability to access the emails in Ireland using computers inside the United States, it declined to comply with the portion of the warrant seeking data stored overseas on the ground that the SCA's mandatory disclosure provisions did not apply extraterritorially.[61]

The district court initially overruled Microsoft's objections, and it held the company in civil contempt for failing to produce the emails.[62] But the U.S. Court of Appeals for the Second Circuit (Second Circuit) reversed

---

vacated and remanded with instructions to dismiss, United States v. Microsoft Corp., No. 17-2, 548 U.S. , 2018 WL 1800369 (U.S. Apr. 17, 2018) (per curiam).

[56] *See supra* § "Mandatory Disclosure Under the SCA."

[57] *See* United States v. Microsoft Corp., No. 17-2, 548 U.S. __, 2018 WL 1800369, slip. op. at 1 (U.S. Apr. 17, 2018) (per curiam).

[58] *Id.*

[59] *Matter of Warrant*, 829 F.3d at 204.

[60] *See* Matter of Warrant, 829 F.3d 197, 204-06 (2d Cir. 2016), *vacated and remanded with instructions to dismiss*, United States v. Microsoft Corp., No. 17-2, 548 U.S. , 2018 WL 1800369 (U.S. Apr. 17, 2018) (per curiam).

[61] *See id.* at 209.

[62] *Id.* at 205.

those rulings in 2016.[63] Relying on the presumption established by the Supreme Court that U.S. laws do not have effect outside U.S. territorial jurisdiction unless the law specifies otherwise,[64] the Second Circuit held that the SCA does not authorize the seizure of emails stored exclusively on foreign servers.[65] The United States appealed the Second Circuit's decision, and the Supreme Court granted certiorari in 2017 in *United States v. Microsoft, Corp.*[66]—a widely followed case that drew attention and amici curie briefs from a range of groups including privacy advocates, law enforcement officials, Members of Congress, 34 U.S. states and territories, and several foreign nations.[67]

The Legislative Response to *Microsoft* in the CLOUD Act

While the *Microsoft* appeal was pending before the Supreme Court, officials from the Department of Justice (DOJ) sought a legislative response to the Second Circuit's ruling.[68] In a hearing before the House Committee on the Judiciary in June 2017,[69] DOJ representatives argued that the Second Circuit's decision "effectively hamstrung the ability of law

---

[63] *See id.* at 222.

[64] *See* RJR Nabisco, Inc. v. European Cmty., 136 S.Ct. 2090, 2101 (2016); Morrison v. Nat'l Australia Bank Ltd., 561 U.S. 247, 266 (2010).

[65] *See Matter of Warrant*, 829 F.3d at 222.

[66] United States v. Microsoft Corp., 138 S.Ct. 356 (2017) (mem. op.), *vacated and remanded with instructions to dismiss*, No. 17-2, 548 U.S. , 2018 WL 1800369 (U.S. Apr. 17, 2018) (per curiam).

[67] Among the more than 30 amici curie briefs were briefs filed by privacy groups; former law enforcement, national security and intelligence officials; 34 U.S. states and territories; the United Kingdom; Ireland; the European Commission (on behalf of the European Union); the New Zealand Privacy Commissioner; two U.S. Senators; and three Members of the U.S. House of Representatives. For a collection of amici briefs filed in *Microsoft*, see United States v. Microsoft Corp., SCOTUSblog (last visited Apr. 19, 2018), http://www.scotusblog.com/case-files/cases/united-states- v-microsoft-corp/.

[68] *See* Legislation to Permit Secure and Privacy-Protected Access to Cross-border Electronic Data for Law Enforcement to Combat Serious Crime and Terrorism [hereinafter 2017 DOJ Proposed Legislation], *in* Downing Statement, *supra* note 1, at app. A. The 2017 DOJ proposal also contained language derived from draft legislation prepared by DOJ in 2016 that addresses authorization for data sharing executive agreements, discussed *infra* § "Executive Agreements Authorized by the CLOUD Act." *See infra* note 174 (discussing the DOJ's legislative proposal in 2016).

[69] *See Data Stored Abroad Hearing*, *supra* note 1.

enforcement" to obtain data stored by U.S. service providers abroad, creating a "tremendous problem" that caused "substantial harm to public safety."[70] Accordingly, DOJ proposed a draft bill that would amend provisions in ECPA, including provisions in the SCA, to state expressly that a service provider must comply with the law's mandatory disclosure requirements when the data is in the provider's possession, custody, or control—regardless of whether the data is located inside the United States.[71] As described by DOJ, the proposal was intended to restore the "pre-*Microsoft* status quo when providers routinely complied" with SCA warrants for data stored abroad.[72]

In February 2018, identical bills—both titled the CLOUD Act—containing DOJ's proposed extraterritoriality provision were introduced in the House and Senate.[73] The CLOUD Act was included in the Consolidated Appropriations Act, 2018, which was passed by both chambers, and signed into law by the President on March 23, 2018.[74] As enacted, the CLOUD Act amends ECPA by, among other things, including the following extraterritoriality provision:

> A [provider] shall comply with the obligations of this chapter to preserve, backup, or disclose the contents of a wire or electronic communication and any record or other information pertaining to a customer or subscriber within such provider's possession, custody, or

---

[70] Downing Statement, *supra* note 1, at 1. *See also* Letter from Samuel R. Ramer, Acting Assistant Att'y Gen., U.S. Dep't of Justice, to the Honorable Paul Ryan, Speaker, U.S. House of Representatives (May 24, 2017), https://judiciary.house.gov/wp-content/ uploads/ 2017/06/Downing-Testimony.pdf [hereinafter Ramer Letter] ("Congress can address the ongoing and substantial damage to public safety caused by the *Microsoft* decision . . . .").

[71] 2017 DOJ Proposed Legislation, *supra* note 68, § 3(a).

[72] Ramer Letter, *supra* note 70, at 1.

[73] *See* H.R. 4943, 115th Cong. (2018); S. 2383, 115th Cong. (2018). The CLOUD Act, as introduced and later enacted into law, contains minor variations on DOJ's proposed extraterritorial provision by removing the reference to a "provider of . . . wire communications"—a term not used in ECPA. *Compare* 2017 DOJ Proposed Legislation, *supra* note 68, § 3(a), *with* CLOUD Act § 103(a)(1) (adding 18 U.S.C. § 2713). The CLOUD Act also added the comity analysis, discussed *infra* § "Resolving Conflicts with Foreign Law," which was not in the 2017 DOJ proposal, and made certain changes to DOJ's proposed authorization for international data sharing agreements, discussed *infra* § "Executive Agreements Authorized by the CLOUD Act."

[74] *See supra* note 11.

> control, regardless of whether such communication, record, or other information is located within or outside of the United States.[75]

After the CLOUD Act's enactment, the United States obtained a new warrant seeking the emails at issue in its dispute with Microsoft under the authority of the new law.[76] Because both the United States and Microsoft agreed that the new warrant replaced the prior warrant, the Supreme Court concluded that the case had become moot, and vacated the lower court's rulings with instructions to dismiss.[77]

Resolving Conflicts with Foreign Law

In addition to defining the extraterritorial reach of the mandatory disclosure provisions in ECPA, including the SCA, the CLOUD Act contains provisions designed to resolve potential conflicts of law that could arise if the United States seeks data stored abroad when the law of a foreign country prohibits disclosure.[78] It does so by authorizing a provider to file a motion to quash or modify a data demand if

- the provider reasonably believes the target of the demand is not a U.S. person[79] and does not reside in the United States;
- the provider reasonably believes disclosure would create a material risk of violating a foreign nation's law; and
- the foreign nation whose law may be violated has a data sharing agreement with the United States authorized by the CLOUD Act (discussed in more detail below[80]).[81]

---

75 CLOUD Act § 103(a)(1) (adding 18 U.S.C. § 2713).

76 United States v. Microsoft Corp., No. 17-2, 548 U.S. __, 2018 WL 1800369, slip op. at 2 (U.S. Apr. 17, 2018) (per curiam).

77 *Id.*

78 CLOUD Act § 103(b) (adding 18 U.S.C. § 2703(h)).

79 The CLOUD Act defines "United States person" as a citizen or national of the United States, an alien lawfully admitted for permanent residence, an unincorporated business association in which a substantial number of members are citizens or lawfully admitted permanent residents, or a corporation that is incorporated in the United States. *See* CLOUD Act § 105(a) (adding 18 U.S.C. § 2523(a)(2)).

A court may grant the providers' motion to modify or quash a government demand for data upon finding that three conditions are met: (1) the required disclosure would violate foreign law; (2) the interests of justice dictate that the demand should be quashed or changed; and (3) the target is not a U.S. person and does not reside in the United States.[82] In determining whether the second condition is satisfied, courts must undertake a "comity analysis."[83] Comity—or respect for foreign sovereignty[84]— is a legal doctrine that, among other things, permits courts to excuse violations of U.S. law, or moderate the sanctions imposed for such violations, when the violations are compelled by a foreign nation's law.[85] Courts and commentators often have described the comity doctrine

---

[80] *See infra* § "Executive Agreements Authorized by the CLOUD Act."

[81] CLOUD Act § 103(b) (adding 18 U.S.C. § 2703(h)). The foreign nation must also provide reciprocal rights allowing providers to quash or modify data demands in the foreign nation. *See id.*

[82] *See id.*

[83] *See id.*

[84] The classic definition of comity in U.S. law is derived from *Hilton v. Guyot*, an 1895 Supreme Court decision:

> "Comity," in the legal sense, is neither a matter of absolute obligation, on the one hand, nor of mere courtesy and good will, upon the other. But it is the recognition which one nation allows within its territory to the legislative, executive or judicial acts of another nation, having due regard both to international duty and convenience, and to the rights of its own citizens, or of other persons who are under the protection of its laws.

159 U.S. 113, 163–64 (1895). For additional background on the comity doctrine, see William S. Dodge, *International Comity in American Law*, 115 Colum. L. Rev. 2071 (2015).

[85] *See* Restatement (Fourth) of Foreign Relations Law: Jurisdiction, Tentative Draft No. 2, § 222 (2016) [Hereinafter Fourth Restatement: Jurisdiction TD 2] ("To the extent permitted by statute, regulation, or procedural rule, U.S. courts have discretion to excuse violations of U.S. law . . . on the ground that the violations are compelled by another state's law, if: (a) the person in question appears likely to suffer severe sanctions for failing to comply with foreign law; and (b) the person in question had acted in good faith to avoid the conflict."); *id.* at § 222 reporters' n.10 (stating that the defense of foreign state compulsion "reflects the practice of states in the interests of comity."). *See also* Société Internationale v. Rogers, 357 U.S. 197, 211 (1958) (ordering lower court to devise less severe sanctions for failure to produce banking records when "the very fact of compliance by disclosure . . . will itself constitute the initial violation of Swiss laws"); Gucci Am., Inc. v. Weixing Li, 768 F.3d 138 (2d Cir. 2014) (directing the district court to "undertake a comity analysis" due to the "apparent conflict between the obligations set forth in [an American court's injunction] and applicable Chinese banking law"); *In re* Sealed Case, 825 F.2d 494, 498 (D.C. Cir. 1987) (reversing dismissal of a contempt order and noting that the "government concedes that it would be impossible for the bank to comply with the contempt order without violating the

as vague and ill-defined,[86] but the CLOUD Act specifically enumerates the factors courts should consider when determining whether comity principles support quashing or modifying a data demand.[87]

Notably, however, the CLOUD Act's comity factors and statutory right to a file a motion to quash or modify apply only to nations with which the United States has a data sharing agreement, as discussed below.[88] For nations with no such agreement, the CLOUD Act preserves common law principles of comity.[89] Common law comity principles generally dictate that U.S. legal obligations can be avoided as a result of foreign law only when the person or entity in question acted in good faith to avoid the conflict, but there remains a likelihood of severe sanctions in the foreign nation for failure to comply with foreign law.[90] Ultimately, the comity analysis under either the CLOUD Act *or* common law principles is likely to be a highly fact-specific evaluation that depends on the specific circumstances of a demand for data stored overseas.

---

laws of country Y on country Y's soil), *cert denied sub nom*, Roe v. United States, 484 U.S. 963 (1987).

[86] *See, e.g.*, JP Morgan Chase Bank v. Altos Hornos de Mexico, S.A. de C.V., 412 F.3d 418, 423 (2d Cir. 2005) ("International comity . . . has never been well-defined."); Turner Entm't Co. v. Degeto Film GmbH, 25 F.3d 1512, 1518 (11th Cir. 1994) (describing "respect for the acts of our fellow sovereign nations" as a "rather vague concept referred to in American jurisprudence as international comity"); Anne-Marie Slaughter, *Court to Court*, 92 Am. J. Int'l L. 708, 708 (1998) ("Comity . . . is a concept with almost as many meanings as sovereignty."); Joel R. Paul, *Comity in International Law*, 32 Harv. Int'l L.J. 1, 4 (1991) ("[D]espite ubiquitous invocation of the doctrine of comity, its meaning is surprisingly elusive.").

[87] The CLOUD Act lists seven factors that the court "shall take into account, as appropriate[,]" in its comity analysis: (A) the United States' interests; (B) the foreign governments' interests; (C) the likelihood, extent, nature and penalties that the provider or its employees could face under foreign law; (D) the location and nationality of the target of the demand, and the nature and extent of the target's connections with the United States and the foreign nation; (E) the nature and extent of the provider's ties to and presence in the United States; (F) the importance of the information to the investigation to be disclosed; (G) the ability to access the information through other means; and (H) the investigative interests of the foreign nation if the data is sought by the United States on behalf of a foreign nation. *See* CLOUD Act § 103(b) (adding 18 U.S.C. § 2703(h)(3)).

[88] *See* CLOUD Act § 103(b) (adding 18 U.S.C. § 2703(h)). *See also* § Executive Agreements Authorized by the CLOUD Act.

[89] *See* CLOUD Act § 103(c).

[90] *See* Fourth Restatement: Jurisdiction TD 2, § 222.

INTERNATIONAL DATA SHARING AFTER THE CLOUD ACT

In addition to expressly expanding the ability of the U.S. government to require service providers to release data stored outside the United States, the CLOUD Act addresses a reciprocal issue: limitations on foreign governments' ability to obtain data in the United States.[91] As internet-based communications have become commonplace, evidence of criminal conduct frequently is derived from data stored on servers located outside the territorial jurisdiction of the nation where the crime was committed.[92] Because technology companies headquartered in the United States hold a majority of the world's electronic communications on their servers, foreign governments frequently seek data held by U.S. companies.[93] At the same time, ECPA prohibits service providers from disclosing the content of electronic communications directly to foreign governments absent a statutory exception or a warrant from a federal court.[94]

With ECPA acting as a "blocking statute" that prevents foreign governments from directly acquiring certain third-party data stored by private entities in the United States, foreign nations have sought the U.S. government's assistance in obtaining warrants that authorize disclosure.[95]

---

[91] *See* CLOUD Act §§ 104-105.

[92] *See supra* notes 1-3. *See also* Letter from Peter J. Kadzik, U.S. Ass't Att'y Gen., to the Hon. Joseph R. Biden, President, U.S. Senate (July 15, 2016), https://tinyurl.com/y7b7fhaw [hereinafter Kadzik Letter] ("Foreign governments investigating criminal activities abroad increasingly require access to electronic evidence from U.S. companies that provide electronic communications to millions of their citizens and residents. Such data is often stored or accessible only in the United States . . . .").

[93] *See* Tiffany Lin and Mailyn Fidler, Cross-Border Data Access Reform: A Primer on the Proposed U.S.- U.K. Agreement 2 (2017), https://dash.harvard.edu/bitstream/handle/1/33867385/2017-09_berklett.pdf?sequence=1 ("Tech companies in the U.S. hold a majority of electronic data, meaning U.K. police investigating a crime in London, for example, may need to access emails stored by a U.S.-based provider."); Woods, *supra* note 1, at 780 ("[T]he vast majority of the world's Internet users store their data with U.S. firms . . . ."); McGuinness Statement, *supra* note 3 ("Most communications services are operated by companies based in the United States.").

[94] *See* 18 U.S.C. § 2702(a)(3).

[95] *See, e.g.*, Aldert Gidari, *The Cross-Border Data Fix: It's Not So Simple*, Center for Internet and Society, Stanford Law School (Jun. 16, 2017) ("[L]aw enforcement outside the U.S. can't get data for their legitimate investigations from U.S. providers because the Electronic Communications Privacy Act (ECPA) prohibits such disclosures; that is, ECPA is a classic

Prior to the CLOUD Act, there were two common international legal processes for obtaining a warrant in the United States: letters rogatory requests and MLATs.[96]

Three Forms of Cross-Border Data Sharing

Letters Rogatory

Discretionary requests made between the courts of one country to the courts of another country that are available to governments and private litigants, which are generally seen as the least efficient and reliable method of obtaining evidence abroad.[97]

Mutual Legal Assistance Treaties (MLATs)

Treaties providing streamlined processes for cross-border evidence sharing between governments in criminal cases, which are reviewed by DOJ and a federal court for compliance with U.S. law.[98]

CLOUD Act Agreements

Executive agreements removing legal restrictions on certain foreign nations' ability to seek data directly from U.S. providers in cases involving "serious crimes" when not targeting U.S. persons, provided that the United States has determined that the foreign nation's laws adequately protect privacy and civil liberties.[99]

Letters Rogatory

Letters rogatory are requests made by a court in one nation to the court of another nation seeking assistance in obtaining evidence located

blocking statute."); *Data Stored Abroad Hearing*, *supra* note 1 (statement of Richard Salgado, Dir. of Law Enforcement and Information Security, Google Inc.), https://judiciary.house.gov/wp- content/uploads/2017/06/Salgado-Testimony.pdf [hereinafter Salgado Statement] ("ECPA includes a broad, so-called 'blocking' provision that restricts the circumstances under which U.S. service providers may disclose the content of users' communications to foreign governments.").

[96] *See* Funk, *supra* note 16, at 1.

[97] *See infra* § "Letters Rogatory."

[98] *See infra* § "Mutual Legal Assistance Treaties (MLATs)."

[99] *See infra* § "Executive Agreements Authorized by the CLOUD Act."

abroad.[100] Historically, letters rogatory were the principle mechanism for sharing evidence between nations.[101] Whereas MLATs and agreements authorized under the CLOUD Act generally are limited to government-to-government requests in criminal cases (with some exceptions in early MLATs),[102] criminal defendants and private litigants in civil cases may request that U.S. courts issue letters rogatory.[103] Governments may also use letters rogatory to seek judicial assistance in obtaining evidence abroad when the United States does not have either an MLAT or a CLOUD Act agreement with a foreign nation.[104]

---

[100] *See* Intel Corp. v. Advanced Micro Devices, Inc., 542 U.S. 241, 248 n.2 (2004) ("[A] *letter rogatory* is the request by a domestic court to a foreign court to take evidence from a certain witness.") (emphasis in original) (quoting Harry Leroy Jones, *International Judicial Assistance: Procedural Chaos and A Program for Reform*, 62 Yale L.J. 515, 519 (1953)); US. Dep't of State, *Preparation of Letters Rogatory*, Travel.State.Gov, https://travel.state.gov/content/travel/en/legal/travel-legal-considerations/internl-judicial-asst/obtaining- evidence/Preparation-Letters-Rogatory.html [hereinafter *Preparation of Letters Rogatory*] ("Letters rogatory are requests from courts in one country to the courts of another country requesting the performance of an act which, if done without the sanction of the foreign court, could constitute a violation of that country's sovereignty.").

[101] *See* Peter Swire & Justin D. Hemmings, *Mutual Legal Assistance in an Era of Globalized Communications: The Analogy to the Visa Waiver Program*, 71 N.Y.U. Ann. Surv. Am. L. 687, 695 (2017) ("[I]nternational information sharing continued to rely on principles of comity and letters rogatory up until 1977.").

[102] While early MLATs entered by the United States allowed criminal defendants to obtain some discovery abroad, more recent treaties expressly state that they do not give rise to a private right to submit requests. *Compare, e.g.*, Mutual Legal Assistance Treaty, arts. 12.2, 18.5, U.S.-Switz., entered into force Jan. 23, 1977, 27 U.S.T. 2019 (permitting criminal defendants or their counsel to be present during the production of witnesses or evidence *In re*sponse to MLAT requests), *with* Agreement on Mutual Legal Assistance, art. 3.5, U.S.-E.U., entered into force Feb. 1, 2010, 43 I.L.M. 758 ("The Contracting Parties agree that this Agreement is intended solely for mutual legal assistance between the States concerned. The provisions of this Agreement shall not give rise to a right on the part of any private person to obtain, suppress, or exclude any evidence, or to impede the execution of a request, nor expand or limit rights otherwise available under domestic law."). *See also* L. Song Richardson, *Convicting the Innocent in Transnational Criminal Cases: A Comparative Institutional Analysis Approach to the Problem*, 26 Berkeley J. Int'l L. 62, 84 (analyzing U.S. MLATs and concluding that all but the three earliest treaties contain clauses restricting defense access to the mutual legal assistance process).

[103] *See, e.g.*, Yonatan L. Moskowitz, *MLATs and the Trusted Nation Club: The Proper Cost of Membership*, 41 Yale J. Intl. L. Online 1, 3 (2016); Funk, *supra* note 16, at 17.

[104] *Preparation of Letters Rogatory*, *supra* note 100 ("Letters rogatory are the customary means of obtaining judicial assistance from overseas in the absence of a treaty or other agreement.").

Letters rogatory are discretionary requests premised on principles of comity rather than an obligation under international law.[105] There is no legal obligation or guarantee that the country receiving the request will respond,[106] and the evidence sharing process has been described as time-consuming and unpredictable.[107] Consequently, letters rogatory are often seen as the least preferable method of obtaining evidence abroad.[108]

Mutual Legal Assistance Treaties (MLATs)

As investigations into complex, coordinated international crimes like money laundering and drug trafficking became more common in the 1970s, the United States and other nations began to enter into MLATs, which established standardized procedures for sharing of certain evidence across national boundaries in criminal matters.[109] MLATs are treaties—

---

[105] *See, e.g.*, *In re* Letters Rogatory from Tokyo Dist., Tokyo, Japan, 539 F.2d 1216, 1219 (9th Cir. 1976) ("[T]he district court is given discretion in determining whether letters rogatory should be honored."); *In re* Letters Rogatory Issued by Na'l Court of First Instance in Commercial Matters N. 23 of Fed. Capital of Argentinean Republic, 144 F.R.D. 272, 274 (E.D. Pa. 1992) ("Because this is a subpoena granted pursuant to Letters Rogatory, this Court has broad discretion to decide whether to honor requests for foreign assistance."); Swire & Hemmings, *supra* note 101, at 692 ("Letters rogatory rely on principles of comity, or respect for foreign sovereignty, rather than on an assertion that the jurisdiction seeking the evidence has a legal right to the evidence."); Funk, *supra* note 16, at 5 (stating that the process for letters rogatory is "more time-consuming and unpredictable" than MLATs "because the enforcement of letters rogatory is a matter of comity between courts, rather than treaty-based").

[106] Funk, *supra* note 16, at 19.

[107] *See, e.g.*, Virginia M. Kendall & T. Markus Funk, *The Role of Mutual Legal Assistance Treaties in Obtaining Foreign Evidence*, 40 Litig. 59, 59 (2014) (describing letters rogatory as "a far less efficient and reliable process" than MLATs); *Preparation of Letters Rogatory*, *supra* note 100 ("Letters rogatory are customarily transmitted via diplomatic channels, a time-consuming means of transmission.").

[108] *See, e.g.*, Office of the United States Attorneys, Criminal Resource Manual § 276, https://www.justice.gov/usam/criminal-resource-manual-276-treaty-requests (describing the MLAT process as "generally faster and more reliable than letters rogatory"); FUNK, *supra* note 16, at 3 ("[P]rosecutors typically consider letters rogatory an option of last resort for accessing evidence abroad, to be exercised only when MLATs are not available"); Woods, *supra* note 1, at 748 (describing letters rogatory as "rarely used and extremely unreliable").

[109] The United States first signed an MLAT with Switzerland in 1973, which entered into force in 1977. *See* Treaty between the United States of America and the Swiss Confederation on Mutual Assistance in Criminal Matters, U.S.- Switz., May 25, 1973, 27 U.S.T. 2019, T.I.A.S. 8302. *See also Consular Conventions, Extradition Treaties, and Treaties Relating*

most often bilateral treaties—in which nations agree to provide certain assistance to foreign governments in the investigation and prosecution of crimes.[110] Whereas letters rogatory are discretionary requests, MLATs create treaty-based obligations governed by international law.[111]

While the requirements in each MLAT may differ depending on the specific terms of the treaty, MLATs generally obligate nations to summon witnesses, compel the production of documents and other evidence, issue warrants, and serve process in response to requests from the foreign government.[112] MLATs typically also identify grounds for refusing requests.[113] The United States has MLATs with more than 60 nations,[114] but this accounts for less than half the nations in the world.[115]

---

to Mutual Legal Assistance in Criminal Matters (MLATs): Hearing Before the S. Comm. on Foreign Relations, 102d Cong. 1, 11 (1992) (statement of Robert S. Mueller, III, Assistant Att'y Gen., Criminal Div., U.S. Dep't of Justice) [hereinafter Mueller Statement] ("We concluded our first MLAT, with Switzerland, to facilitate access to Swiss bank records. Financial records are vital to the successful prosecution of organized crime bosses and drug kingpins, who are rarely caught red-handed . . . ."); Richardson, *supra* note 102, at 98 (providing background on the U.S.-Swiss MLAT).

[110] For a list of U.S. MLATs, see 2 U.S. Dep't of State, Bureau for Int'l Narcotics and Law Enforcement Affairs, International Narcotics Control Strategy Report: Money Laundering and Financial Crimes 21 (2014)[hereinafter Strategy Report] and 7 Foreign Affairs Manual (F.A.M.) § 962.1(d), https://fam.state.gov/FAM/ 07FAM/07FAM0960.html.

[111] *See In re* Commissioner's Subpoena, 325 F.3d 1287, 1292–1304 (11th Cir. 2003) (explaining that "[l]aw enforcement authorities found the statute" authorizing federal district courts to entertain letters rogatory "to be an unattractive option in practice because it provided wide discretion in the district court to refuse the request and did not obligate other nations to return the favor that it grants. MLATs, on the other hand, have the desired quality of compulsion as they contractually obligate the two countries to provide to each other evidence and other forms of assistance needed in criminal cases while streamlining and enhancing the effectiveness of the process for obtaining needed evidence."), *abrogated in part on other grounds by* Intel Corp. v. Advanced Micro Devices, Inc., 542 U.S. 241 (2004); Swire & Hemmings, *supra* note 101, at 695-96 (describing the development of comity-based requests to treaty- based requests).

[112] 7 F.A.M. § 962.1(a). *See also* Funk, *supra* note 16, at 5 (listing common types of assistance in MLATs).

[113] *See, e.g.*, Treaty Between the United States and Ukraine on Mutual Legal Assistance in Criminal Matters, U.S.-Ukr., art. 3, entered into force Feb. 27, 2001, S. Treaty Doc. 106-16 (stating that the central authority of the requesting state may deny assistance if, among other reasons, the request relates to an offense under military law or would prejudice the "security or similar essential interests" of the receiving state).

[114] The United States has bilateral MLATs with more than 50 nations and is also a party to the multilateral Agreement on Mutual Legal Assistance with the European Union and the Inter-American Convention on Mutual Legal Assistance of the Organization of American States. *See* Strategy Report *supra* note 110, at 21. The United States is also a party to other multilateral treaties, such as the International Convention for the Suppression of the

Each party to an MLAT designates a central authority through which direct communications can be made.[116] The central authority for the United States is the Office of International Affairs (OIA) in the Criminal Division of DOJ.[117] When a request for legal assistance is submitted to the United States,[118] OIA receives and conducts an initial review to ensure that the request contains all necessary information and comports with required formats.[119] OIA then transmits the request to the U.S. Attorney in the jurisdiction where the witness or evidence is located.[120] The U.S. Attorney brings the request before a federal district court by filing a request for a court order or warrant authorizing the United States to carry out the action sought by the foreign nation.[121] Before authorizing the action, courts review the request to ensure that it complies with the underlying treaty and U.S. law and constitutional requirements.[122] After a warrant or court order

---

Financing of Terrorism, *opened for signature* Jan. 10, 2000, 2178 U.N.T.S. 197, and the United Nations Convention Against Corruption, *opened for signature* Dec. 9, 2003, 2349 U.N.T.S. 41, which provide for cooperation in the investigation and prosecution of the particular offenses that are the subject of the treaties. *See id*; Restatement (Fourth) of Foreign Relations Law: Jurisdiction, Tentative Draft No. 3, § 313 reporters n.1 (2017).

[115] *See* U.S. Dep't of State, Bureau of Intelligence and Research, *Independent States in the World* (Jan. 20, 2017), https://www.state.gov/s/inr/rls/4250.htm (identifying 195 independent nations). *See also* Downing Statement, *supra* note 1, at 7 ("[T]he United States maintains bilateral MLA treaties with less than one-half of the world's countries.").

[116] 7 F.A.M. § 962.1(a); Mueller Statement, *supra* note 109, at 11 ("The most significant benefit of MLATs may lie in institutionalizing law enforcement cooperation . . . by mandating for each treaty partner a Central Authority which serves as the clearinghouse for all incoming and outgoing requests.").

[117] 7 F.A.M. § 962.1(c).

[118] Outgoing MLAT requests from the United States to foreign nations often follow similar procedures as incoming requests, but the process depends on the nation receiving the request. *See* Bitkower Statement, *supra* note 4, at 21 (discussing the general procedure through which OIA serves MLAT requests on foreign nations); Swire et al., *supra* note 4, at 357 (detailing the process by which the United States submits MLAT requests to France).

[119] *See* Swire & Hemmings, *supra* note 101, at 698. For additional background the MLAT process, see Funk, *supra* note 16, at 6-11.

[120] There are 93 U.S. Attorneys stationed throughout the United States and its territories, and each serves as the "chief federal law enforcement officer of the United States within his or her particular jurisdiction." U.S. Dep't of Justice, Office of the Attorney General, *Mission*, Justice.gov (last updated Sep. 22, 2016), https://www.justice.gov/usao/mission.

[121] *See* Funk, *supra* note 16, at 6; Swire & Hemmings, *supra* note 101, at 699.

[122] *See In re* Dolours Price, 685 F.3d 1, 15 (1st Cir. 2012) ("It is undisputed that treaty obligations are subject to some constitutional limits."); *In re* Premises Located at 840 140th Avenue NE, Bellevue, Washington, 634 F.3d 557, 572 (9th Cir. 2011) ("At a minimum, the Constitution requires that a request not be honored if the sought-after information would be used in a foreign judicial proceeding that 'depart[s] from our concepts of fundamental due

has been issued and the provider transfers the data to the U.S. government, OIA and the Federal Bureau of Investigation (FBI) review the material in an effort to minimize production of information that is not responsive to the request.[123]

According to the 2013 President's Review Group on Intelligence and Communications Technologies, MLAT requests submitted to the United States take an average of approximately 10 months to complete.[124] When the United States seeks data from foreign nations, some requests take "considerably longer,"[125] especially when submitted to countries that are uncooperative or have less sophisticated legal systems.[126] According to one U.S. official, the United States never receives a response to some requests.[127]

Executive Agreements Authorized by the CLOUD Act

Although the MLAT process generally is seen as more predictable and efficient than letters rogatory,[128] MLATs became the subject of criticism in recent years due to, among other things, the typical length of response time under such agreements and the fact that the United States does not have any MLAT with more than half the nations in the world.[129] At the same time, the number of requests for assistance in obtaining data and other evidence in the United States has increased markedly. In its FY2017

---

process and fairness.'") (quoting *In re* Request for Judicial Assistance from Seoul District Criminal Court, 555 F.2d 720, 724 (9th Cir. 1977)); Funk, *supra* note 16, at 5 ("[T]he district court must still review the terms of each request, checking that they comply with the terms of the underlying treaty and comport with U.S. law.").

[123] *See* Swire & Hemmings, *supra* note 101, at 699.

[124] *See* President's Review Group, *supra* note 17, at 227.

[125] *See* Bitkower Statement, *supra* note 4, at 21.

[126] *Id.*

[127] *Id.*

[128] *See supra* note 107-108.

[129] *See, e.g.*, President's Review Group, *supra* note 17, at 227 (identifying problems with and proposing six steps to improve the MLAT process); Bitkower Statement, *supra* note 4, at 35-36; Gail Kent, *The Mutual Legal Assistance Problem Explained*, CTR. for Internet and Soc'y, Stanford Law Sch. (Feb. 23, 2015), http://cyberlaw.stanford.edu/blog/2015/02/mutual-legal-assistance-problem-explained. *See also supra* note 114 (discussing the nations with which the U.S. has MLATs).

budget request, DOJ stated that the number of requests for judicial assistance from foreign countries increased nearly 85%, and the number for requests for "computer records" increased over 1000%.[130]

As foreign governments' need for data located overseas has expanded, some nations have sought data directly from U.S. providers and passed legislation authorizing their governments to compel disclosure.[131] These developments have placed U.S. technology companies at the intersection of potentially conflicting legal obligations: service providers may be both subject to foreign court orders compelling the release of data and prohibited by U.S. law from disclosing that data.[132] The potentially conflicting obligations coupled with criticisms of the MLAT and letters rogatory processes led to proposals for changes in the international data sharing regime that ultimately culminated in the CLOUD Act.[133]

The CLOUD Act creates a third paradigm of international data sharing arrangements: the possibility of international agreements that remove legal restrictions on U.S. technology companies' ability to disclose data directly to certain foreign nations in response to "orders" issued by foreign nations.[134] Whereas MLATs are "treaties" within the meaning of U.S. constitutional law—meaning they are binding international agreements concluded by the Executive after receiving the advice and consent of the

---

[130] Criminal Div., U.S. Dep't of Justice, Performance Budget: FY 2017 President's Budget 23 (2016), http://www.justice.gov/jmd/file/820926/download.

[131] *See* Downing Statement, *supra* note 1, at 8. *See also* Jonah Force Hill, *Problematic Alternatives: MLAT Reform for the Digital Age*, Harv. Nat'l Sec. L. J. (Jan. 28, 2015), http://harvardnsj.org/2015/01/problematic-alternatives-mlat- reform-for-the-digital-age/ (discussing foreign nations' desire to obtain data from U.S. companies through foreign subsidiaries).

[132] *See* Downing Statement, *supra* note 1, at 8 ("Our companies may face conflicting legal obligations when foreign governments require them to disclose electronic data in the United States that U.S. law prohibits them from disclosing"); Smith Statement, *supra* note 3, at 62 (describing conflicting legal obligations faced by Microsoft as result of Brazilian court orders compelling the disclosure of the contents of electronic communications stored outside Brazil).

[133] *See* CLOUD Act § 102 (including in congressional findings that "[t]imely access to electronic data held by communications-service providers is an essential component of government efforts to protect public safety and combat serious crime," but that such access is "impeded by the inability to access data stored outside the United States[,]" and potentially subject to "conflicting legal obligations" under U.S. and foreign law).

[134] *See* CLOUD Act §§ 104-105.

Senate as provided in the Treaty Clause[135]—the CLOUD Act authorizes the United States to enter "executive agreements" with qualifying foreign nations.[136] Executive agreements are binding international agreements entered into by the Executive based on a source of authority other than the Treaty Clause.[137] The Executive's authority often is derived from legislation, as is the case in the CLOUD Act.[138]

The executive agreements authorized under the CLOUD Act would allow service providers to disclose the contents of electronic communications—both stored communications and real-time communications intercepted by wiretap—directly to requesting foreign governments with whom the United States has an authorized data sharing agreement.[139] The act does so by removing ECPA's prohibitions on disclosure to such foreign governments.[140] When a foreign nation with a CLOUD Act agreement issues an "order" seeking data from a provider in the United States, the provider can deliver the requested data without civil

---

[135] *See* U.S. Const., art. II, § 2, cl. 2 ("The President . . . shall have Power, by and with the Advice and Consent of the Senate, to make Treaties, provided two thirds of the Senators present concur[.]"). The term "treaty" has a broader meaning under international law, in which it is generally synonymous with all binding agreements, than in the context of domestic law, in which it refers to the subcategory of international agreements that are concluded by the President after receiving the advice and consent of the Senate. *See* CRS Report RL32528, *International Law and Agreements: Their Effect upon U.S. Law*, by Michael John Garcia, at 2.

[136] CLOUD Act § 105.

[137] Although not mentioned expressly in the Constitution, the executive branch has entered into executive agreements on a variety of subjects without the advice and consent of the Senate since the early years of the Republic. *See, e.g.*, Am. Ins. Ass'n v. Garamendi, 539 U.S. 396, 415 (2003) ("[O]ur cases have recognized that the President has authority to make 'executive agreements with other countries, requiring no ratification by the Senate . . . this power having been exercised since the early years of the Republic"); L. Henkin, Foreign Affairs and the United States Constitution 219 (2d ed. 1996) ("Presidents . . . have made many thousands of [executive] agreements, differing in formality and importance, on matters running the gamut of U.S. foreign relations."). For additional background on the difference between treaties and executive agreements, see CRS Report RL32528, *supra* note 135, at 2-9.

[138] Executive agreements that are authorized by legislation enacted through the bicameral process are known as "congressional-executive" agreements. *See* CRS Report RL32528, *supra* note 135, at 5.

[139] *See* CLOUD Act § 104.

[140] The CLOUD Act amends portions of the Wiretap Act (18 U.S.C. §§ 2511(2), 2520(d)), the SCA (*id.* § 2702(b)-(c)), and the Pen Register Statute (*id.* §§ 3121(a), 3124(d)) by permitting disclosure pursuant to an executive agreement authorized by the act. *See* CLOUD Act § 104.

or criminal penalty under ECPA.[141] By contrast, in the MLAT and letters rogatory processes, cross-border data requests initially are submitted to government entities rather than to the private party in possession of the data.[142]

Although the CLOUD Act authorizes executive agreements that would remove ECPA's prohibitions on disclosure, neither the act nor the agreements it authorizes create a legal obligation for service providers to comply with foreign governments' data demands.[143] Rather, a foreign government's authority to issue an order seeking data must derive solely from its domestic law.[144] Additionally, state or federal laws other than ECPA still may prohibit disclosure of particular classes of information.[145]

Requirements for CLOUD Act Agreements

The CLOUD Act contains a number of restrictions on the type of foreign governments with whom the United States can enter agreements and the nature of demands for data that qualifying foreign governments can issue to U.S. providers.[146] Before an agreement concluded under the CLOUD Act can enter into force, the Attorney General, with the concurrence of the Secretary of State, must make four written certifications that are provided to Congress and published in the *Federal Register*:

1. the foreign nation's domestic law "affords robust substantive and procedural protections for privacy and civil liberties" in its data-

---

[141] In addition to removing prohibitions in the Wiretap Act, SCA, and Pen Register statute, *supra* note 140, the CLOUD Act amends each act to make a good faith belief that disclosure was permitted pursuant to an executive agreement a defense to liability. *See* CLOUD Act § 104.

[142] *See supra* §§ "Letters Rogatory"; "Mutual Legal Assistance Treaties (MLATs)."

[143] CLOUD Act § 105 (requiring that "any obligation for a provider of electronic communications service or remote computing service to produce data" under a CLOUD Act agreement "shall derive solely" from the foreign nation's law).

[144] *Id.*

[145] *See, e.g.*, 12 U.S.C. § 3402 (providing "no Government authority may have access to or obtain copies of, or the information contained in the financial records of any customer from a financial institution unless" statutory exceptions apply); 18 U.S.C. § 2710 (restricting disclosure of "prerecorded video cassette tapes or similar audio visual materials").

[146] *See* CLOUD Act § 105.

collection activities, as determined based on at least seven statutory factors;[147]
2. the foreign government has adopted "appropriate" procedures to minimize the acquisition, retention, and dissemination of information concerning U.S. persons;
3. the executive agreement will not create an obligation that providers be capable of decrypting data, nor will it create a limitation that prevents providers from decryption;[148] and
4. the executive agreement will require that any order issued under its terms will be subject to an additional set of procedural and substantive requirements, as discussed below.[149]

The CLOUD Act expressly states that these certifications are not subject to judicial or administrative review.[150] But the act gives Congress the power to prevent a proposed executive agreement from entering into force through expedited congressional review provisions after the certifications are provided.[151] Certifications must be renewed every five years, and recertifications trigger Congress's power to block renewal

[147] The CLOUD Act provides that the factors "to be met" when determining whether a foreign government affords the requisite protections for privacy and civil liberties include the following: whether the foreign government (1) has "adequate" laws related to cybercrime and electronic evidence as demonstrated by being a party to the Convention on Cybercrime, entered into force Jan. 7, 2004, 41 I.L.M. 282, 2296 U.N.T.S. 167 (known as the Budapest Convention) or through domestic law consistent chapters I and II of the Budapest Convention; (2) demonstrates "respect for rule of law and principles of nondiscrimination;" (3) "adheres to international human rights obligations and commitments or demonstrates respect for international universal human rights[;]" (4) "has clear legal mandates and procedures" governing its entities that are authorized to seek data, including procedures through which those authorities "collect, retain, use, and share data, and effective oversight of those activities;" (5) has "sufficient mechanisms to provide accountability and appropriate transparency regarding the collection and use of electronic data[;]" and (6) "demonstrates a commitment to promote and protect the global free flow of information and the open, distributed, and interconnected nature of the Internet . . . ." *See* CLOUD Act § 105.

[148] For background on decryption, see CRS Report R44642, *Encryption: Frequently Asked Questions*, by Chris Jaikaran, at 2.

[149] *See* CLOUD Act § 105 (adding 18 U.S.C. § 1253).

[150] *Id.* ("A determination or certification made by the Attorney General . . . shall not be subject to judicial or administrative review.").

[151] The procedures for expedited review in Congress are discussed *infra* § "Congressional Review of CLOUD Act Agreements."

through expedited review processes.[152] Additionally, if requested by the Committees on the Judiciary or Foreign Affairs in the House or the Committees on the Judiciary or Foreign Relations in the Senate, the executive branch must furnish to the requesting committee a summary of the factors it considered when determining that a foreign government satisfies the CLOUD Act's requirements.[153]

Limitations on Orders Issued Under CLOUD Act Agreements

The fourth certification required by the CLOUD Act mandates that any data sharing agreement concluded under the act contain a set of requirements related to foreign governments' orders issued to service providers. These include, among things,[154] requirements that all orders

- identify a specific person, account, or other identifier that is the object of the order;[155]
- be premised on a "reasonable justification based on articulable and credible facts, particularity, and severity regarding the conduct under investigation";[156]
- not intentionally target a U.S. person (or person located in the U.S.) or target a non-U.S. person with the intention of obtaining information about a U.S. person;
- be issued for the purpose of obtaining information relating to the prevention, detection, investigation, or prosecution or a "serious "crime"—a term that the CLOUD Act states includes terrorism, but otherwise does not define;[157]
- comply with the domestic law of the issuing country;
- not be used to infringe freedom of speech; and

---

[152] *See* CLOUD Act § 105 (adding 18 U.S.C. § 1253).

[153] The CLOUD Act requires that a proposed agreement and the Attorney General's certifications be transmitted to the Committees on the Judiciary and Foreign Affairs in the House of Representatives and the Committees on the Judiciary and Foreign Relations in the Senate. *See id.*

[154] The description of requirements for CLOUD Act agreements in the body of this chapter is not exhaustive. A complete list of requirements is contained in Section 105 of the act.

[155] *See* CLOUD Act § 105 (adding 18 U.S.C. § 1253).

[156] *See id.*

[157] *See id.*

- satisfy additional requirements for real-time communications captured by wiretap.[158]

When a foreign government receives the requested data from the provider, it must promptly review the material and store any unviewed communications on a "secure system accessible only to those trained in applicable procedures . . . ."[159] The "applicable procedures" must, to the maximum extent possible, comply with the minimization procedures in Section 101 of the Foreign Intelligence Surveillance Act (FISA).[160] Foreign governments may not issue an order at the request of the United States or any third-party government, and they may not disclose the content of communications of a U.S. person to the U.S. government except in cases involving significant harm or threat of harm to the United States or U.S. persons.[161]

Mandatory Rights Granted to the United States

The CLOUD Act requires that data sharing agreements grant certain powers to the U.S. government. Specifically, the foreign government must grant reciprocal rights of data access to the United States and allow the U.S. government to conduct periodic reviews of the foreign nation's compliance with the terms of the executive agreement.[162] CLOUD Act agreements also must reserve the United States' right to "render the

---

[158] Wiretap orders must be for a fixed, limitation duration; may not last longer than is reasonably necessary to accomplish the purposes of the order; and can be issued only if the information could not be obtained with less intrusive methods. *See id.*

[159] *Id.*

[160] See 50 U.S.C. § 1801(h). For background on FISA and its minimization procedures, see CRS Report R44457, *Surveillance of Foreigners Outside the United States Under Section 702 of the Foreign Intelligence Surveillance Act (FISA)*, by Edward C. Liu, at 2-4, and Congressional Distribution Memorandum from Edward C. Liu, Legislative Attorney, Cong. Research Serv., Summary of Substantive Provisions of S. 2010, the FISA Amendments Reauthorization Act of 2017, H.R. 3989, the USA Liberty Act of 2017, and S. 139, the FISA Amendments Reauthorization Act of 2017, at 7-17 (available to congressional clients from the author upon request).

[161] *See* CLOUD Act § 105 (adding 18 U.S.C. § 1253).

[162] *See id.*

agreement inapplicable" for any order for which the United States concludes the agreement "may not properly be invoked."[163]

Judicial or Governmental Review of Orders Under CLOUD Act Agreements

The process for judicial or other government oversight of foreign nations' requests for data under the CLOUD Act differs from earlier international data sharing regimes. In both the MLAT and letters rogatory processes, a federal court reviews and approves a foreign government's request for information before issuing a warrant or court order.[164] Such requests generally must satisfy U.S. legal standards and constitutional requirements, such as the Fourth Amendment probable cause standard.[165] Several federal appellate courts have stated that an otherwise valid MLAT or letters rogatory request may be rejected if compliance would result in a violation of the Constitution.[166] For MLAT requests, agencies in the executive branch conduct additional reviews for compliance with U.S. law before and after receiving judicial approval to execute a cross- border data request.[167]

Under CLOUD Act agreements, by contrast, foreign governments can submit orders directly on service providers.[168] While those orders are "subject to review or oversight by a court, judge, magistrate, or other independent authority" in the *foreign nation*, the CLOUD Act does not require review or approval by a U.S. court or federal agency.[169] And unlike MLATs and letters rogatory, the CLOUD Act contemplates that the judicial or other independent review in the foreign country could occur

---

[163] *Id.*

[164] *See* Funk, *supra* note 16, at 10-11, 18-19.

[165] *See* Kendall & Funk, *supra* note 107, at 60 ("[Federal judges . . . serve as the gatekeepers for search warrants, wiretaps, and other methods of obtaining evidence, ensuring that the requested foreign evidence collection meets the same standards as those required in U.S. cases . . . for example, finding probable cause . . . ."); Woods, *supra* note 1, at 783 ("Under the current ECPA regime, foreign law enforcement officials must prove to a U.S. judge that they have probable cause (the Fourth Amendment standard) to obtain a warrant.").

[166] *See supra* note 122.

[167] *See* Swire & Hemmings, *supra* note 101, at 696-700.

[168] *See* CLOUD Act § 104.

[169] *Id.* § 105 (adding 18 U.S.C. § 1253).

after a foreign government issued an order to a service provider.[170] The ultimate result is that foreign nations' orders issued under the CLOUD Act are not required to undergo individualized review by any branch of the U.S. government, and U.S. courts are not required to analyze whether the foreign government's request complies with U.S. constitutional standards. This change appears to be intended to accelerate the data sharing process, especially in cases involving emergency or other time-sensitive requests.[171] Rather than review each request individually, the United States' opportunity to scrutinize a foreign country's data demands primarily will occur during the periodic review of a foreign nation's compliance with its data sharing agreements and when evaluating whether a foreign nation's laws satisfy the CLOUD Act's eligibility requirements.[172]

What Nations Are Eligible for CLOUD Act Agreements?

The CLOUD Act does not specify by name what countries meet its requirements, and the Attorney General has not provided the requisite certifications for a proposed agreement as of the date of this chapter. Consequently, it is not clear which, if any, nations may be eligible for CLOUD Act agreements. However, in 2016, DOJ informed Congress that the United States sought legislation that would implement a potential bilateral data sharing agreement with the United Kingdom.[173] While the draft bilateral agreement has not been made public, DOJ proposed legislation that the department stated was necessary to implement the

[170] *See id.* (providing that judicial or independent review must take place "prior to, *or* in proceedings regarding, enforcement of the order . . . .") (emphasis added).

[171] *See, e.g.*, Downing Statement, *supra* note 1, at 9 (contending that legislative reform to the MLAT process is necessary to allow more expedient access to digital evidence); McGuinness Statement, *supra* note 3 (same).

[172] *Cf.* Lin & Fidler, *supra* note 93, at 5 ("[O]rders do not undergo individual inspection by the U.S. government, making the vetting of countries for the executive agreement the single guaranteed point of scrutiny.").

[173] *See* Kadzik Letter, *supra* note 92 ("The legislative proposal is necessary to implement potential bilateral agreement between the United Kingdom and the United States that would permit U.S. companies to provide data In response to U.K. orders targeting non-U.S. persons located outside the United States, while affording the United States reciprocal rights . . . .").

potential agreement.[174] The structure and many provisions of the CLOUD Act appear to have been derived—and in some cases taken verbatim—from DOJ's proposed legislation.[175] Some commentators believe that the U.S.-U.K. agreement will be the first agreement to be certified by the executive branch and submitted to Congress for review under the CLOUD Act's expedited congressional review procedures, as discussed below.[176]

Congressional Review of CLOUD Act Agreements

The CLOUD Act provides for a mandatory 180-day period of congressional review before a proposed data sharing agreement can enter into force.[177] The act also defines a number of procedures authorizing congressional consideration of a joint resolution of disapproval of an executive agreement on an expedited process. The procedures include among other things, automatic discharge of the congressional committees to whom the joint resolution has been referred within 120 days;[178] waiver of certain points of order; limitations on and structuring of debate; and

---

[174] *See* Legislation to Permit the Secure and Privacy-Protective Exchange for Electronic Data for the Purposes of Combating Serious Crime Including Terrorism [hereinafter 2016 Proposed U.S.-U.K. Legislation] *in* Kadzik Letter, *supra* note 92.

[175] *Compare, e.g.*, 2016 Proposed U.S.-U.K. Legislation, *supra* note 174, § 2(1) ("Timely access to electronic data held by communications-service providers is an essential component of government efforts to protect public safety and combat serious crime, including terrorism . . . ."), *with* CLOUD Act § 102(1) (identical language). DOJ proposed amending ECPA to add an extraterritoriality provision in response to *Microsoft* in a draft bill circulated in 2017. *See supra* note 68. That 2017 proposal incorporated the provisions authorizing data sharing executive agreements from DOJ's 2016 proposal. *See id.*

[176] *See, e.g.*, Thomas P. Bossert & Paddy McGuinness, Opinion, *Don't Let Criminals Hide Their Data Overseas*, N.Y. Times (Feb. 15, 2018), https://www.nytimes.com/2018/02/14/opinion/data-overseas-legislation.html ("The bill would authorize the attorney general to enter into such agreements, but only with allies that respect privacy and protect civil liberties, and that have records of promoting and defending due process. The first one would be with Britain, which already has the authority to enter into such a pact."); Jennifer Daskal, *New Bill Would Moot Microsoft Ireland Case— And Much More!*, Just Security (Feb. 6, 2018), https://www.justsecurity.org/51886/bill-moot-microsoft-ireland-case- more/ ("[T]he legislation would authorize the executive to finalize a draft executive agreement with the UK that was negotiated during the Obama presidency . . . .").

[177] CLOUD Act § 105 (adding 18 U.S.C. § 1253).

[178] A joint resolution of disapproval is automatically referred to the House Committees on the Judiciary and Foreign Affairs and the Senate Committees on the Judiciary and Foreign Relations. *Id.* Whereas Congress's 180-day period to vote on a joint resolution of disapproval commences on the date on which the Attorney General provides a copy of the proposed agreement to Congress, the 120-day clock for committee consideration begins to run on the date of referral of a joint resolution. *Id.*

expedited treatment of a joint resolution received from the other chamber of Congress.[179]

If Congress enacts a joint resolution of disapproval during the 180-day review window, the CLOUD Act states that the proposed agreement may not enter into force.[180] Such a joint resolution of disapproval would require passage by both chambers of Congress and the President's signature or a veto override.[181] Because the CLOUD Act provides that proposed data sharing agreements will be submitted to Congress after already receiving the approval of two Cabinet-level executive officials—the Attorney General and Secretary of State—some commentators contend that a President would be unlikely to sign a joint resolution of disapproval, making a veto-proof majority necessary to block a proposed CLOUD Act agreement.[182]

Commentary on the CLOUD Act

The CLOUD Act has garnered both praise and criticism from observers.[183] Some argue that the act provides a practical remedy for problems related to the globalization of evidence and the increased demand for data stored overseas in criminal cases.[184] Supporters assert that the need for data stored abroad, which often is held by U.S. internet companies, has overburdened the legal architecture established in the MLAT and letters

---

[179] *See id.*

[180] *See id.*

[181] *See Legislation, Laws, and Acts*, U.S. Senate (last visited Apr. 5, 2018), https://tinyurl.com/yaun8wry ("Like a bill, a joint resolution requires the approval of both Chambers in identical form and the president's signature to become law. There is no real difference between a joint resolution and a bill.").

[182] *See, e.g.*, Neema Singh Gullani & Naureen Shah, *The CLOUD Act Doesn't Help Privacy and Human Rights: It Hurts Them*, Lawfare (Mar. 16, 2018), https://lawfareblog.com/cloud-act-doesnt-help-privacy-and-human-rights-it- hurts-them; Robyn Greene, *Four Common Sense Fixes to the CLOUD Act that its Sponsors Should Support*, Just Security (Mar. 13, 2018), https://www.justsecurity.org/53728/common-sense-fixes-cloud-act-sponsors-support/.

[183] *See infra* notes 184-190.

[184] *See, e.g.*, Bossert & McGuinness, *supra* note 176; Lisa Monaco & John P. Carlin, Opinion, *A "Global Game of Whack-a-Mole": Overseas Data Rules are Stuck in the 19th Century*, Wash. Post (Mar. 5, 2018), https://tinyurl.com/ybghkrhn; Andrew Keane Woods, Peter Swire, *The CLOUD Act: A Welcome Legislative Fix for Cross-Border Data Problems*, Lawfare (Feb. 6, 2018), https://lawfareblog.com/cloud-act-welcome-legislative-fix- cross-border-data-problems.

rogatory systems, rendering those systems "outdated and inefficient."[185] Supporters also argue that the CLOUD Act provides adequate protection for privacy, civil liberties, and human rights.[186] They contend that, absent the change in law, frustrated foreign governments that are unable to obtain data held by U.S. companies will exert extraterritorial application of their own laws or enact data localization laws[187] that some believe impede the effective functioning of an open internet.[188] Several major U.S. technology companies—including Apple, Facebook, Google, Microsoft, and Oath—support the legislation, calling it an effective legislative solution that reduces conflicts of laws.[189]

Critics of the CLOUD Act argue that it poses a threat to civil liberties and human rights by lowering the standards previously necessary to obtain evidence in cross-border criminal investigations and prosecutions.[190] They contend that the CLOUD Act's standard for individualized suspicion—

---

[185] *See* Lin & Fidler, *supra* note 93, at 4.

[186] *See, e.g.*, Jennifer Daskal, Peter Swire, *Why the CLOUD Act is Good for Privacy and Human Rights*, Lawfare (Mar. 14, 2018), https://www.lawfareblog.com/why-cloud-act-good-privacy-and-human-rights.

[187] Data localization laws require technology companies to store data on servers within nations' respective borders, thereby potentially obviating the need for cross-border data requests. *See, e.g.*, Bret Cohen, Britanie Hall, Charlie Wood, *Data Localization Laws and Their Impact on Privacy, Data Security and the Global Economy*, Antitrust, Fall 2017, at 107 ("Russia, China, Indonesia, and others have enacted explicit 'forced' localization requirements applicable to broad swaths of industry that require data to be stored on servers within their respective borders . . . ."); William Alan Reinsch, *A Data-Localization Free-for-all?*, Center for Strategic & International Studies (Mar. 9, 2018), https://www.csis.org/blogs/future-digital-trade-policy-and-role-us-and-uk/data-localization-free-all#_ednref1 ("The degree of data localization measures worldwide has increased dramatically, most drastically since 2010."). For a survey of global data localization measures, see Anupam Chander & Uyên P. Lê, *Data Nationalism*, 64 Emory L.J. 677, 682-712 (2015).

[188] *See, e.g.*, Lin & Fidler, *supra* note 93, at 4; Jennifer Daskal, Peter Swire, *Privacy and Civil Liberties Under the CLOUD Act: A Response*, Lawfare (Mar. 21, 2018), https://www.lawfareblog.com/privacy-and-civil-liberties-under- cloud-act-response.

[189] *See* Letter from Apple et al. to Representative Doug Collins et al. (Feb. 6, 2018), https://blogs.microsoft.com/datalaw/wp-content/uploads/sites/149/2018/02/Tech-Companies-Letter-of-Support-for- House-CLOUD-Act-020618.pdf.

[190] *See, e.g.*, Sharon Bradford Franklin, Director of Surveillance & Cybersecurity Policy, New America, Open Technology Institute, *OTI Opposes the CLOUD Act*, Open Technology Institute (Feb. 6, 2018), https://www.newamerica.org/oti/press-releases/oti-opposes-cloud-act/; Gullani & Shah, *supra* note 182; Robyn Greene, *Somewhat Improved, the CLOUD Act Still Poses a Threat to Privacy and Human Rights*, Just Security (Mar. 23, 2018), https://www.justsecurity.org/54242/improved-cloud-act-poses-threat-privacy-human-rights/.

"reasonable justification based on articulable and credible facts, particularity, legality, and severity regarding the conduct under investigation"—is vague and may not rise to the level of probable cause necessary to obtain a judicial warrant under U.S. law.[191] Some argue that the executive branch's decision to certify a country as satisfying the CLOUD Act's standards should be subject to judicial or other review.[192] Others contend that the concept that foreign nations' data requests do not need individualized review if the nations' domestic laws meet the act's eligibility criteria is flawed because foreign governments' real-world operations may not comport with their domestic laws and may change over time.[193] Several critics of the CLOUD Act argue that it should require a foreign court or independent authority to approve a foreign government's order before the order is issued on a U.S. provider.[194] Others contend, among other things, that the law should increase the requirements for foreign governments to obtain access to real-time communications to the same standards that apply to the United States' interception of live communications in the Wiretap Act.[195]

How Will CLOUD Act Agreements Interact with Existing Data Sharing Processes?

Executive agreements authorized by the CLOUD Act would supplement, not replace, existing avenues of international data sharing.[196]

---

[191] *See* Gullani & Shah, *supra* note 182. *See also* Franklin *supra* note 190; Camille Fischer, *The CLOUD Act: A Dangerous Expansion of Snooping on Cross-Border Data*, Electronic Frontier Foundation (Feb. 8, 2018), https://www.eff.org/deeplinks/2018/02/cloud-act-dangerous-expansion-police-snooping-cross-border-data; *CLOUD Act Would Erode Trust in Privacy of Cloud Storage*, Center for Democracy and Technology (Feb. 6, 2018), https://cdt.org/press/cloud-act-would-erode-trust-in-privacy-of-cloud-storage/.

[192] *See, e.g.*, Franklin *supra* note 190.

[193] *See* Gullani & Shah, *supra* note 182 ("The very premise of the current CLOUD Act—the idea that countries can effectively be safe-listed as human-rights compliant, such that their individual data requests need no further human rights vetting—is wrong.").

[194] *See, e.g.*, Daniel Sepulveda, Opinion, *Bill on Cross-Border Data Access Needs to Change, Despite Laudable Goal*, The Hill (Mar. 16, 2018), http://thehill.com/opinion/technology/378785-bill-on-cross-border-data-access-needs-to- change-despite-laudable-goal; Greene, *supra* note 190; Franklin *supra* note 190.

[195] *See* Fischer, *supra* note 191; Greene, *supra* note 190; Gullani & Shah, *supra* note 182.

[196] *See* CLOUD Act § 106.

Accordingly, requests for assistance would still be available through MLATs (when in effect) and letters rogatory.

When analyzed in light of existing data sharing processes, the CLOUD Act has the potential to result in a three-tiered system for cross-border data sharing in criminal matters. Those nations that are approved for CLOUD Act agreements could request data directly from U.S. service providers in cases involving "serious crimes"—provided they do not target U.S. persons or persons located in the United States and meet the CLOUD Act's other requirements.[197] For nations that have an MLAT but no CLOUD Act agreement, or for data requests that fall outside the scope of the CLOUD Act, foreign governments can use the MLAT process.[198] Finally, private litigants and nations that do not have a CLOUD Act agreement or an MLAT may request that their courts issue letters rogatory to the courts of the United States.[199]

CONCLUSION

While the CLOUD Act is likely to more clearly define the scope of U.S. officials' right to seek certain data stored overseas in the custody of U.S. providers, its broader impact on the international data sharing regime is less certain. As the internet continues to expand and become more globalized, law enforcement officials worldwide can be expected to continue to seek access to data stored on servers outside their territorial jurisdictions.[200] Although the major technology companies responsible for maintaining a large share of the world's data are located in the United States,[201] the United States accounts for less than 10% of the estimated 3

---

[197] *See supra* § "Requirements for CLOUD Act Agreements."

[198] *See supra* § "Mutual Legal Assistance Treaties (MLATs)."

[199] *See supra* § "Letters Rogatory."

[200] *See, e.g.*, Woods, *supra* note 1, at 741-42 (discussing shifts in expansion of internet usage across the globe); *Chapter One Cooperation or Resistance?: The Role of Tech Companies in Government Surveillance*, 131 Harv. L. Rev. 1722 (2018) ("[T]echnology companies have become major actors in the world of law enforcement and national security.").

[201] *See supra* note 93.

billion internet users worldwide.[202] These demographics potentially could lead many nations to pursue CLOUD Act agreements, which would provide faster access to data held by U.S. providers. Whether the United States ultimately enters such agreements will depend on the willingness of the executive branch to certify foreign nations' eligibility and Congress's desire to block a proposed agreement through a joint resolution of disapproval enacted into law.

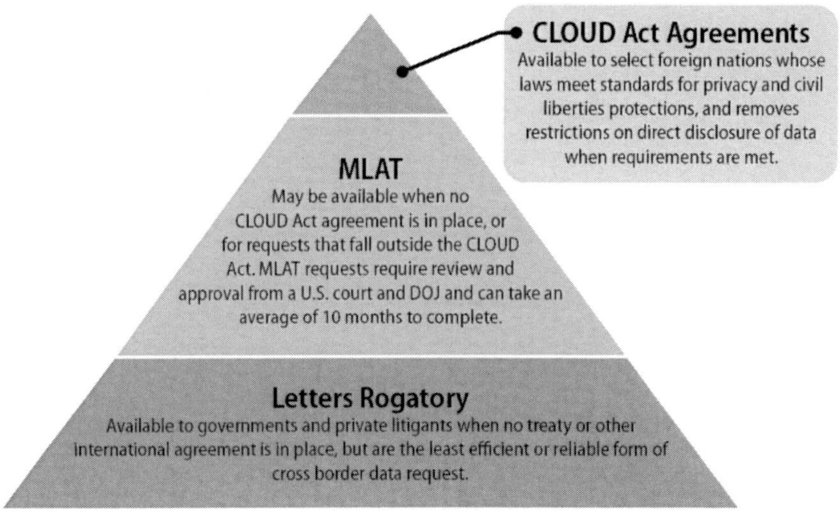

Source: *Supra* §§ "Letters Rogatory"; "Mutual Legal Assistance Treaties (MLATs)"; and "Executive Agreements Authorized by the CLOUD Act."

Figure 1.Three Tiers of Cross-Border Data Sharing.

The impact of the CLOUD Act on privacy, human rights, and civil liberties interests similarly is difficult to predict.[203] The act has the potential to create a three-tiered system of international data sharing, with the United States' most trusted foreign partners able to obtain data directly from U.S.

---

202 Woods, *supra* note 1, at 741.

203 *Cf.* Tom Kulik, *Stormy Weather: How the Cloud Act May Rain on the Privacy of Data*, Above the Law (Apr. 13, 2018), https://tinyurl.com/y82ze95b ("[T]he Cloud Act has definitely created some unpredictable weather . . . .").

companies without individualized review by the U.S. government.[204] Because this system of direct access differs from existing international data sharing regimes, the manner in which data requests are administered, the type of data that is collected, and the degree of potential for abuse of the system, if any, may become more apparent over time.

---

[204] *Cf.* Moskowitz, *supra* note 103, at 2 (discussing the potential formation of a so-called "Trusted Nations Club" in the context of international data sharing); Swire and Hemmings, *supra* note 101, at 690 (analogizing a cross-border data sharing regime to the Visa Waiver Program in which citizens of a group of developed nations can bypass certain requirements for travel to the United States).

In: New Strategies in Cloud Computing
Editor: Luc Larivière
ISBN: 978-1-53616-907-2

Chapter 6

SUPREME COURT TO HEAR DIGITAL PRIVACY CASE: CAN THE GOVERNMENT OBTAIN EMAILS STORED ABROAD?*

Austin D. Smith

Update: In United States v. Microsoft Corp., the Supreme Court was set to address whether a warrant issued under the Stored Communications Act (SCA) allowed U.S. law enforcement to compel Microsoft to hand over emails it stores on a server overseas. Following the publication of this Sidebar, and less than one month after the Court heard oral argument in Microsoft, Congress passed and the President signed into law the Clarifying Lawful Overseas Use of Data (CLOUD) Act as part of the Consolidated Appropriations Act, 2018. The CLOUD Act amended the SCA to require that technology companies provide data in their possession, custody, or control in response to an SCA warrant—regardless of whether the data is located in the United States. The government then obtained a new warrant pursuant to the CLOUD Act and the Supreme Court dismissed the case as moot.

* This is an edited, reformatted and augmented version of Congressional Research Service Publication No. LSB10086, dated February 23, 2018.

The original Sidebar post previewing the Microsoft case, published on February 23, 2018, is below.

Can U.S. law enforcement use a warrant to compel Microsoft to hand over emails it stores on a server in Ireland? The Supreme Court may soon answer that question in a case, United States v. Microsoft, that could have major repercussions for both digital privacy rights and law enforcement interests. The case comes on appeal from a 2016 decision by the U.S. Court of Appeals for the Second Circuit (Second Circuit), discussed in a previous Legal Sidebar, which held that the government could not use the Stored Communications Act (SCA) to compel Microsoft to disclose emails it stored abroad. While both the government and Microsoft agree that the SCA does not apply overseas, the central disagreement before the Supreme Court is whether using the SCA to allow the government to access electronic communications held by a domestic entity but stored abroad would constitute a domestic or foreign application of the law. This Legal Sidebar provides background on the case, including an overview of the SCA, discusses the specific issues presented by the Microsoft case, and explores the case's implications for Congress.

THE STORED COMMUNICATIONS ACT AND EXTRATERRITORIALITY

Enacted in 1986, the SCA governs when and how a provider of electronic communications or remote computing services may disclose communications it stores. The statute generally prohibits providers of electronic communications from disclosing those communications to third parties. However, in a provision at issue in *Microsoft,* the statute requires disclosure to the government pursuant to a warrant based on probable cause that the communications contain evidence of a crime. Importantly, the SCA is silent on whether the law applies only in the United States or globally.

That silence is significant because longstanding precedent establishes a presumption that United States laws do not have any effect outside the country unless they specify otherwise. This interpretive rule of thumb protects against unintended conflicts between American and foreign laws that could lead to unnecessary international incidents. When this presumption applies, a critical question for courts then becomes whether a law meant only for the territorial United States is impermissibly being applied abroad or whether the law's application is appropriately domestic in nature.

Prior Supreme Court cases have laid out the test for whether an application of a law is domestic or extraterritorial: courts must look to the "'focus' of congressional concern"—what conduct the law regulates and who the law protects. The Supreme Court has explained that "[i]f the conduct relevant to the statute's focus occurred in the United States, then the case involves a permissible domestic application even if other conduct occurred abroad; but if the conduct relevant to the focus occurred in a foreign country, then the case involves an impermissible extraterritorial application regardless of any other conduct that occurred in U. S. territory." The Supreme Court has not, however, addressed what the "focus of congressional concern" was with regard to the SCA.

United States v. Microsoft

In this vein, the *Microsoft* litigation centers on whether a particular application of the SCA is impermissibly extraterritorial in nature. The case results from how Microsoft stores its customers' emails as part of its web-based email service. The company stores each individual user's emails in one of its many datacenters around the world—generally the one closest to where a user states that she is from when she signs up for the service. This case began when the government suspected that a user of Microsoft's service had communicated about illegal drug trafficking over email. The government obtained a warrant for the emails in the usual way: by satisfying a magistrate judge that there was probable cause that a crime had been committed and that the emails hosted by Microsoft were evidence of that crime. It then served the warrant on Microsoft and demanded the emails and related information.

While Microsoft complied with the warrant as to the metadata about the user that it stored on its servers located in the United States, it refused to turn over the contents of the emails, which are stored on a Microsoft server in Dublin, Ireland. Microsoft admits that it has the ability to access the emails from its U.S. computers, but contends that the SCA does not extend extraterritorially and, therefore, does not authorize the government to obtain emails that are housed on servers overseas.

The government and Microsoft agree that the SCA does not apply extraterritorially. However, they disagree over what the "focus" of the SCA is and therefore over whether this is a case that involves an impermissible extraterritorial application of that law. The government argues that the focus of the relevant provision of the SCA is on *disclosures* of electronic communications. Because the disclosure in this case would occur in the United States when Microsoft gives the emails to the government, the government maintains that this case presents a permissible domestic application of the law. In other words, for the government, the location of the emails is irrelevant because disclosure, as opposed to access to information by a provider, is at the heart of the SCA. The government further argues that even if the focus of the SCA is on user privacy, any privacy invasion occurs when the data is transferred to the government in the United States, not when the information is first transferred from abroad Microsoft contends instead that the focus of the SCA should be inferred from the statute's overall structure, not just the single provision that allows the government to require disclosures in certain circumstances. Microsoft thus argues, in line with the Second Circuit's holding, that the statute's focus is on the security of communications in domestic electronic storage. Because the company would have to retrieve emails stored in Ireland to comply with the warrant, Microsoft argues that the government's warrant is a result of an impermissible extraterritorial application of the law.

Ramifications and Issues for Congress

The ultimate ruling in *Microsoft* will likely be important regardless of which side prevails. A decision affirming the Second Circuit would likely make it more difficult for federal or state prosecutors to access electronic communications held by American companies if those communications are stored abroad. As the government notes, under Microsoft's theory a domestic company could even avoid complying with the SCA entirely by moving all of its servers overseas. If the government cannot use the SCA to obtain communications stored on servers in foreign countries, it would likely have to resort to processes defined in Mutual Legal Assistance Treaties (MLATs), which generally enable multinational cooperative law enforcement activities. However, the United States has MLATs with fewer than half of the nations of the world (though it does with Ireland), and the process has been criticized as "antiquated and slow." Some scholars have suggested that a ruling for Microsoft could encourage efforts to streamline the MLAT process and incentivize the execution of such treaties with more nations.

On the other hand, Microsoft argues that a ruling for the government that would allow U.S. law enforcement to access communications stored overseas could potentially create conflicts for service providers because many other nations regulate when and under what circumstances data stored within their borders can be transmitted to other countries. The European Union (EU), for example, filed an amicus brief laying out the basic contours of EU laws governing the transfer of personal data to non-EU states, which are quite stringent. Some companies argue conflicts could arise between U.S. and foreign law that would necessitate a violation of the laws of at least one of the jurisdictions the companies are subject to if the government prevails in this case. A win for the government might also embolden other countries to attempt to seize data stored here: Microsoft's brief poses the hypothetical of a Chinese warrant demanding disclosure of a journalist's emails from a U.S. server, and also points to the testimony of the company's president and chief legal officer before the House Judiciary Committee regarding real- world conflicts that have arisen in recent years from Brazil attempting to unilaterally force Microsoft to disclose emails the company stores in the United States.

Two Senators and three Representatives have filed an amicus brief in *Microsoft,* arguing that because "Congress did not intend or expect the SCA to authorize the seizure of data held within the territory of a foreign, sovereign nation," the Second Circuit's decision should be affirmed and

Congress should be allowed to determine the answer to this issue legislatively. To that end, they, along with other cosponsors, have introduced the Clarifying Lawful Overseas Use of Data (CLOUD) Act—S. 2383; H.R. 4943—an amendment to the SCA that potentially could moot the *Microsoft* case by specifying how communications stored abroad may be accessed by domestic law enforcement. (The CLOUD Act would also set up a regime for allowing foreign governments to access communications stored in the U.S. in certain circumstances.) The CLOUD Act is an extension of the earlier proposed International Communications Privacy Act—S. 1671; H.R. 3718—discussed in the amicus brief.

Beyond the specific question at issue in *Microsoft*, the SCA has more broadly been the subject of criticism, with one leading commentator describing the Act as "dense and confusing." Because the SCA is structured around the state of technology as it existed in the mid-1980s, as the *Microsoft* litigation demonstrates, applying the law to modern situations can be fairly difficult. As this chapter notes, Members of Congress have proposed various efforts to update the 1986 law in recent years. The Supreme Court's decision in *Microsoft* may clarify the international scope of the SCA and, with that, could inform the extent to which Congress may decide to amend the law.

INDEX

A

access, vii, viii, x, xi, xii, xiii, 2, 5, 12, 14, 45, 49, 57, 105, 107, 108, 109, 110, 111, 115, 117, 122, 124, 147, 152, 153, 161, 163, 169, 172, 175, 176, 179, 182, 185, 198, 199, 200, 201, 202, 203, 206, 207, 208, 209, 214, 215, 217, 219, 222, 224, 227, 229,230, 233, 234, 236, 238, 240, 241, 242
acquisitions, viii, 3, 10, 19, 21, 24, 29, 41, 43, 51, 52, 56, 58, 113, 158, 165, 170, 191
agencies, vii, viii, 2, 3, 4, 5, 6, 7, 8, 9, 10, 11, 12, 14, 15, 16, 17, 18, 19, 20, 21, 22, 23, 24, 25, 26, 27, 28, 29, 30, 31, 32, 33, 34, 35, 36, 37, 38, 39, 40, 41, 43, 44, 45, 46, 47, 48, 49, 50, 51, 52, 53, 54, 55, 56, 57, 61, 62, 64, 65, 66, 67, 68, 69, 70,71, 72, 73, 74, 75, 76, 77, 78, 79, 80, 101, 106, 107, 108, 113, 116, 118, 145, 146, 147, 148, 149, 150, 151, 152, 153, 154, 155, 156, 157, 158, 159, 160, 161, 162, 164, 165, 206, 228
agility, 72, 104, 106, 111, 119, 121, 187
automate, 46, 103, 123, 178, 189
automation, 103, 115, 123, 175
avoidance, 37, 39, 57, 64
awareness, 48, 153, 163, 191

B

benefits, vii, viii, ix, xiii, 2, 4, 5, 8, 9, 10, 12, 14, 25, 43, 44, 49, 50, 56, 63, 64, 66, 71, 72, 73, 101, 108, 109, 110, 111, 149, 165, 168, 183, 187, 191, 204
Beyond Cloud First, viii, 146
bilateral, 219, 220, 229
breakdown, 26, 30, 31, 37
business processes, 46, 123
businesses, 47

C

candidates, 25, 157, 162, 163, 164
certification, 47, 54, 161, 162, 225, 226
challenges, 11, 19, 34, 40, 102, 104, 109, 111, 113, 154, 155, 156, 162, 169, 172, 174, 175, 178, 179, 180, 184, 201

civil liberties, xi, xii, 200, 201, 202, 203, 216, 224, 225, 230, 232, 235
Clarifying Lawful Overseas Use of Data (CLOUD), v, vii, viii, ix, x, xiii, 1, 2, 5, 6, 7, 12, 14, 15, 17, 18, 19, 20, 21, 22, 23, 25, 28, 29, 30, 35, 40, 41, 43, 44, 45, 46, 47, 48, 50, 52, 53, 55, 56, 57, 62, 66, 68, 72, 74, 75, 76, 77, 78, 79, 80, 81, 82, 83, 84, 85, 86, 87, 88, 89, 90, 91, 92, 93, 94, 95, 96, 97, 98, 99, 100, 101, 102, 103, 104, 105, 106, 107, 108, 109, 111, 112, 113, 114, 116, 117, 119, 120, 121, 122, 123, 145, 146, 147, 148, 150, 151, 156, 157, 159, 164, 165, 167, 170, 171, 172, 174, 175, 177, 178, 180, 181, 184, 185, 186,187, 188, 189, 190, 193, 194, 195, 196, 197, 198, 199, 200, 201, 202, 203, 204, 208, 210, 211, 212, 213, 214, 215, 216, 217, 221, 222, 223, 224, 225, 226, 227, 228, 229, 230, 231, 232, 233, 234, 235, 237, 242
CLOUD Act, v, viii, x, xiii, 199, 200, 201, 202, 203, 204, 208, 210, 211, 212, 213, 214, 215, 216, 217, 221, 222, 223, 224, 225, 226, 227, 228, 229, 230, 231, 232, 233, 234, 235, 237, 242
cloud computing, vii, viii, ix, 2, 3, 4, 5, 8, 11, 12, 14, 15, 16, 17, 19, 22, 24, 27, 32, 33, 36, 37, 40, 41, 44, 49, 51, 54, 57, 58, 59, 60, 61, 62, 64, 68, 69, 71, 73, 74, 76, 77, 101, 147, 155, 156, 161, 162, 164, 165, 169, 170, 173, 176, 177, 178, 181, 183, 185, 188, 191, 193, 194, 198, 201
Cloud computing, vii, viii, 2, 44, 77, 178, 201
Cloud First, v, viii, 5, 7, 20, 22, 56, 68, 145, 190
cloud investments, vii, viii, 2, 3, 4, 5, 8, 26, 27, 28, 29, 30, 34, 36, 37, 38, 56, 66, 70, 72, 77
cloud services, vii, viii, ix, 2, 3, 4, 5, 6, 7, 8, 9, 11, 13, 14, 15, 17, 18, 19, 20, 21, 22, 23, 24, 25, 26, 27, 28, 29, 30, 31, 32, 33, 34, 35, 36, 37, 38, 39, 40, 41, 42, 43, 44, 45, 46, 47, 48, 49, 50, 52, 53, 54, 55, 56, 57, 58, 59, 60, 61, 62, 63, 64, 66, 67, 68,69, 71, 72, 73, 78, 80, 101, 102, 124, 154, 156, 158, 170, 174, 180, 193, 194, 195, 196, 198
cloud smart, v, vii, viii, 17, 40, 57, 145, 146, 147, 148, 150, 151, 156, 157, 159, 164, 165, 180, 181, 185
cloud-based solutions, vii, viii, 145, 146, 156
commercial, 11, 41, 47, 54, 121, 123, 124, 153, 156, 157, 158, 159, 169, 170, 173, 174, 175, 176, 178, 180, 181, 182, 184, 186, 187, 188, 194, 195, 196, 197, 203
Communications Act, x, xi, xiii, 200, 202, 204, 207, 237, 238
compliance, 43, 47, 69, 104, 150, 151, 158, 170, 213, 216, 227, 228, 229
computer, 4, 162, 169, 193, 201, 205, 208, 222
computer systems, 4
computing, vii, viii, ix, 2, 3, 4, 5, 8, 11, 12, 13, 14, 15, 16, 17, 19, 22, 24, 27, 32, 33, 36, 37, 40, 41, 44, 49, 51, 54, 57, 58, 59, 60, 61, 62, 64, 68, 69, 71, 73, 74, 76, 101, 110, 112, 147, 155, 156, 159, 161, 162, 164, 165, 168, 169, 170, 172, 173, 176, 177, 178, 179, 181, 183, 185, 187, 188, 191, 193, 194, 195, 196, 198, 202, 205, 206, 207, 224, 238
Congress, iv, vii, x, xii, xiii, xiv, 10, 11, 19, 40, 41, 43, 64, 197, 198, 199, 202, 208, 210, 211, 224, 225, 229, 230, 231, 235, 237, 238, 240, 241, 242
cost, vii, viii, 2, 4, 5, 9, 10, 12, 14, 16, 19, 20, 22, 25, 29, 34, 35, 36, 37, 38, 39, 41, 42, 44, 45, 46, 57, 58, 59, 60, 61, 64, 66, 70, 71, 72, 102, 104, 110, 112, 114, 115, 119, 121, 148, 154, 156, 169, 170, 173, 179

cost saving, 5, 10, 25, 29, 35, 37, 38, 39, 66, 114, 148
crimes, x, xi, xii, 200, 201, 202, 203, 216, 218, 234
criminal investigations, 201, 203, 232
critical infrastructure, 184
customer service, 4, 47, 54
customers, x, 13, 48, 106, 147, 153, 175, 199, 205, 209, 239
cyberattack, 203
cybersecurity, 48, 118, 148, 153, 160, 162, 164, 165, 174, 175, 178
cyberspace, 168

D

data center, x, xi, xii, 14, 29, 35, 38, 40, 42, 46, 53, 108, 111, 112, 114, 169, 175, 179, 183, 185, 199, 200, 202, 209
data collection, 7, 8, 35, 39, 70, 71
database, 11, 102, 104, 105, 106
decision makers, 107, 122, 168, 173, 176, 184
Department of Agriculture, 1, 6, 24, 27, 31, 32, 36, 50, 67, 77, 79, 80, 116
Department of Commerce, 1, 6, 24, 27, 31, 32, 36, 50, 67, 77, 79, 83, 119, 125
Department of Defense (DoD), v, vii, ix, 1, 6, 24, 27, 31, 32, 36, 50, 67, 77, 79, 84, 85, 121, 142, 167, 168, 169, 170, 171, 172, 173, 174, 175, 176, 177, 178, 179, 180, 181, 182, 183, 184, 185, 186, 187, 188, 189, 190, 191, 193
Department of Education, 1, 6, 24, 27, 31, 32, 36, 67, 77, 79, 87, 127
Department of Energy, 1, 6, 24, 27, 31, 32, 36, 67, 77, 79, 88, 128
Department of Health and Human Services, 2, 6, 24, 27, 31, 32, 36, 67, 77, 79, 89, 90, 92, 93
Department of Homeland Security, 1, 6, 24, 27, 31, 32, 36, 50, 67, 75, 77, 79, 93, 113, 132, 152
Department of Justice, 2, 6, 24, 27, 31, 32, 36, 50, 67, 77, 79, 95, 111, 210
Department of Labor, 2, 6, 24, 27, 31, 32, 36, 62, 67, 77, 79, 97, 139
Department of the Treasury, 2, 6, 24, 27, 31, 32, 36, 67, 77, 79, 97, 102
Department of Transportation, 2, 6, 24, 27, 31, 32, 36, 50, 67, 77, 79, 98, 99, 104, 136
Departments of Agriculture, 5, 44, 51, 66
digital evidence, 229
digital world, vii, ix, 167
disclosure, x, xi, xii, 153, 200, 202, 203, 204, 205, 206, 207, 209, 211, 212, 213, 215, 222, 223, 224, 238, 240, 241
distributed computing, 180
draft, 9, 17, 61, 63, 73, 75, 76, 165, 196, 210, 211, 229, 230
drug trafficking, 218
due process, 221, 230

E

economies of scale, 170, 186
ecosystem, ix, 168
education, 46, 148, 161, 206
educational institutions, 118
electronic communications, vii, x, xi, xiii, 199, 200, 202, 204, 205, 209, 215, 222, 223, 224, 238, 240
emergency, 114, 206, 229
emergency response, 114
employees, 108, 161, 162, 163, 164, 214
encryption, 152, 174, 175, 189
enforcement, x, xi, xiii, 157, 199, 200, 209, 215, 218, 219, 220, 229, 238
environment, ix, 14, 54, 62, 106, 112, 114, 147, 160, 161, 164, 168, 172, 174, 175,

176, 177, 179, 180, 182, 183, 185, 187, 188, 189, 190, 198
Environmental Protection Agency, 5, 66
European Commission, 210
European Union, 210, 219, 241
evidence, x, xi, xii, 9, 73, 199, 200, 201, 202, 207, 215, 216, 217, 218, 219, 220, 221, 225, 228, 231, 232, 238, 239
execution, viii, 146, 159, 173, 175, 177, 178, 217, 241
executive branch, xi, xii, 200, 203, 223, 226, 228, 230, 233, 235

F

faith, 206, 213, 214, 224
federal agency, 19, 228
federal cloud computing strategy, viii, 5, 14, 16, 17, 20, 75, 145, 196
Federal Emergency Management Agency, 93, 94
federal funds, 206
federal government, 4, 9, 10, 16, 19, 43, 74, 107, 109
Federal Highway Administration, 99
federal law, 195, 204, 209, 220, 224
Federal Register, 224
Federal Reserve, 102, 103
Federal Student Aid, 88
financial, 47, 53, 56, 104, 108, 122, 148, 173, 189, 224
financial records, 224
first responders, 115
fiscal year, viii, 2, 3, 5, 6, 7, 15, 16, 17, 20, 23, 24, 25, 26, 27, 28, 29, 30, 31, 32, 33, 34, 36, 37, 40, 42, 46, 47, 57, 66, 67, 68, 69, 70, 77, 78, 79, 80, 111, 124
force, 169, 177, 178, 217, 218, 219, 224, 225, 230, 231, 241
funding, 43, 46, 52, 56, 115
funds, 11, 41, 43, 47, 198

G

global infrastructure, vii, ix, 167
global scale, 185
globalization, 201, 231
governance, 10, 43, 50, 51, 52, 122, 150, 153, 158, 173, 185, 187, 188, 190, 191
governments, x, xi, xii, 199, 200, 201, 202, 203, 206, 214, 215, 216, 219, 222, 223, 224, 226, 227, 228, 232, 233, 234, 242
guidance, viii, 2, 3, 4, 7, 13, 14, 15, 16, 18, 19, 20, 21, 22, 23, 25, 28, 29, 30, 33, 34, 35, 39, 56, 57, 58, 59, 60, 61, 62, 64, 67, 68, 69, 72, 75, 76, 78, 146, 148, 155, 156, 157, 159, 161, 165, 170, 171, 190, 194, 195
guidelines, 175
guiding principles, ix, 167, 180

H

House of Representatives, 65, 66, 210, 211, 226
Housing and Urban Development, 5, 11, 66
human, xi, 12, 52, 164, 179, 200, 202, 225, 231, 232, 233, 235
human capital, 52, 164
human right, xi, 200, 202, 225, 232, 233, 235

I

illegal drug trafficking, 209, 239
improvements, 30, 41, 43, 44, 103, 157, 159
individuals, 55, 109, 151, 164
industry, ix, 157, 160, 162, 163, 169, 173, 181, 186, 193, 196, 197, 203, 232
information sharing, 4, 153, 155, 203, 217
information technology, 2, 3, 4, 10, 16, 66, 169

infrastructure, vii, ix, 13, 28, 42, 44, 45, 46, 48, 54, 76, 113, 114, 119, 121, 147, 149, 150, 159, 167, 168, 169, 170, 171, 173, 174, 176, 178, 180, 183, 184, 187, 188, 191, 194, 195
investments, vii, viii, ix, 2, 3, 4, 5, 6, 7, 8, 9, 10, 18, 19, 20, 21, 22, 23, 24, 25, 26, 27, 28, 29, 30, 31, 32, 33, 34, 36, 37, 38, 39, 40, 43, 49, 56, 57, 58, 59, 60, 61, 62, 64, 66, 68, 69, 70, 71, 72, 73, 77, 78, 80, 146, 165
issues, xiii, xiv, 4, 7, 34, 40, 42, 51, 70, 75, 112, 115, 118, 122, 154, 204, 223, 238

J

job satisfaction, 162
Joint Enterprise Defense Infrastructure (JEDI), v, vii, ix, 186, 187, 193, 194, 195, 196, 197, 198
jurisdiction, x, 200, 201, 207, 210, 215, 218, 220
justification, 226, 233

L

law enforcement, xi, xiii, 48, 201, 205, 210, 211, 220, 228, 234, 237, 238, 241, 242
laws, xi, xii, 10, 200, 203, 210, 213, 216, 225, 229, 232, 233, 238, 241
leadership, 19, 43, 51, 56, 119, 148, 150, 155, 161, 162, 164, 190
learning, 148, 182, 183
legislation, 210, 222, 223, 229, 230, 232
litigation, xiii, 203, 239, 242
local government, 14, 115, 118

M

majority, 3, 20, 26, 29, 33, 169, 186, 215, 231
management, 10, 11, 12, 15, 20, 29, 34, 35, 37, 39, 41, 43, 46, 47, 49, 50, 51, 52, 53, 55, 63, 107, 112, 116, 117, 119, 121, 146, 147, 149, 150, 153, 154, 156, 157, 158, 164, 174, 175, 179
marketplace, 154, 156, 157, 188, 197
materials, 53, 206, 224
media, x, xi, 163, 199, 200, 204
Medicaid, 89, 90
medical, 47, 110
Medicare, 89, 90, 93
Microsoft, viii, x, xii, xiii, 196, 200, 201, 202, 203, 204, 208, 209, 210, 211, 212, 222, 230, 232, 237, 238, 239, 240, 241, 242
migration, 4, 8, 11, 14, 16, 25, 28, 29, 35, 39, 42, 44, 56, 57, 58, 59, 60, 61, 64, 72, 74, 75, 76, 81, 148, 153, 160, 161, 164, 165, 170, 178, 185, 187, 188, 189, 190
military, vii, ix, 121, 124, 167, 168, 171, 175, 177, 180, 191, 219
military exercises, 124
mission, ix, 8, 9, 17, 44, 45, 48, 49, 55, 72, 73, 102, 104, 108, 118, 121, 147, 148, 150, 154, 158, 161, 165, 167, 169, 171, 173, 177, 178, 180, 187, 188, 189, 220
mobile device, 12, 111
modernization, ix, xi, xii, 11, 30, 41, 42, 43, 74, 76, 146, 148, 149, 150, 165, 170, 190, 200, 203
money laundering, 218

N

national security, 168, 172, 210, 234
natural disasters, 115, 118

O

Office of Management and Budget (OMB), vii, viii, 2, 3, 4, 5, 6, 7, 10, 14, 15, 16, 17, 18, 20, 22, 23, 25, 28, 29, 30, 33, 34, 35, 37, 38, 39, 40, 41, 51, 56, 57, 58, 59, 60, 61, 64, 65, 66, 67, 68, 69, 70, 72, 74, 75, 76, 78, 80, 111, 116, 145, 151, 153, 155, 157, 161, 165
officials, viii, x, xi, xii, 2, 6, 7, 8, 21, 22, 23, 24, 25, 26, 27, 28, 29, 33, 37, 38, 39, 41, 42, 44, 45, 46, 47, 48, 49, 50, 52, 53, 54, 55, 56, 63, 68, 70, 71, 72, 76, 102, 103, 104, 105, 106, 107, 108, 109, 110, 111, 112, 113, 114, 115, 116, 117, 118, 119, 120, 121, 122, 123, 124, 197, 199, 200, 203, 209, 210, 228, 231, 234
on-demand access, vii, viii, 2, 12
operating system, 13, 42, 45
operations, 10, 15, 19, 29, 43, 44, 50, 52, 55, 72, 73, 102, 103, 106, 107, 108, 111, 114, 116, 120, 121, 122, 124, 153, 168, 171, 173, 177, 178, 181, 194, 233
opportunities, 19, 54, 108, 150, 155, 160, 161, 162, 164, 175, 177, 179
optimization, 26, 175
oversight, 10, 17, 19, 20, 34, 41, 51, 52, 225, 228

P

platform, 11, 42, 46, 55, 106, 109, 183, 188
policy, 5, 17, 20, 22, 75, 76, 148, 151, 152, 153, 165, 232
President, viii, xii, xiii, 16, 48, 74, 145, 150, 201, 202, 203, 211, 215, 221, 222, 223, 231, 237
principles, 156, 157, 159, 197, 214, 217, 218, 225
private party, 224
private sector, viii, 15, 145, 156, 162, 165
procedural rule, 213
procurement, vii, viii, 19, 22, 51, 52, 81, 146, 149, 155, 156, 158, 161, 165, 169
professional development, 161, 163
professionals, 155, 156, 161, 162, 164, 165, 183
protection, ix, 152, 153, 168, 206, 207, 213, 232
public safety, 211, 222, 230
public-private partnerships, 160

R

recommendations, iv, viii, 3, 16, 17, 18, 51, 57, 61, 62, 63, 65, 74, 76, 145, 155
recovery, 104, 114, 115, 156, 178
reliability, 7, 8, 70, 71, 104
requirements, ix, xi, xii, 10, 13, 14, 38, 39, 48, 49, 51, 52, 53, 55, 113, 116, 118, 148, 150, 151, 152, 153, 154, 159, 160, 161, 165, 171, 175, 177, 181, 191, 193, 194, 197, 198, 200, 203, 205, 207, 208, 211, 219, 220, 225, 226, 227, 228, 229, 232, 233, 234, 236
resources, vii, viii, 2, 5, 12, 13, 43, 44, 45, 48, 49, 50, 51, 52, 53, 54, 55, 102, 104, 107, 108, 109, 110, 113, 115, 119, 120, 121, 146, 148, 149, 160, 161, 162, 165, 168, 169, 170, 173, 174, 189, 194
response, x, xii, xiii, 17, 61, 113, 115, 170, 196, 200, 202, 210, 217, 219, 221, 222, 229, 230, 232, 237
responsiveness, 72, 104, 105, 106, 111, 119, 121
restrictions, xi, xii, 200, 203, 204, 216, 222, 224
rights, iv, xiii, 206, 213, 217, 225, 227, 229, 231, 232, 233, 238
risk, 10, 22, 55, 115, 148, 149, 150, 151, 152, 154, 155, 156, 158, 159, 175, 180, 181, 189, 212

risk assessment, 155
risk management, 148, 154, 156, 158

S

savings, vii, viii, 2, 3, 4, 7, 10, 29, 35, 36, 37, 38, 39, 40, 41, 42, 43, 56, 57, 58, 59, 60, 61, 64, 70, 71, 80, 101, 156
Secretary of Agriculture, 57
Secretary of Commerce, 57
Secretary of Defense, ix, 58, 168, 191, 193, 198
Secretary of Homeland Security, 59
Secretary of the Treasury, 60
security, vii, viii, 1, 2, 4, 5, 6, 14, 15, 24, 27, 31, 32, 36, 43, 44, 48, 49, 50, 51, 54, 55, 59, 61, 66, 67, 75, 76, 77, 79, 83, 86, 89, 90, 93, 94, 95, 97, 100, 101, 102, 103, 104, 113, 115, 116, 123, 132, 141, 146, 149, 150, 151, 152, 153, 154, 155, 156, 158,159, 165, 168, 169, 170, 171, 172, 174, 175, 176, 177, 179, 181, 188, 189, 195, 203, 210, 216, 219, 230, 231, 232, 234, 240
security practices, 155
security services, 189
servers, x, xi, 5, 12, 13, 45, 194, 199, 200, 201, 210, 215, 232, 234, 239, 241
service provider, xii, 12, 13, 14, 22, 42, 49, 53, 55, 152, 153, 154, 157, 158, 202, 208, 211, 215, 216, 222, 223, 224, 226, 228, 230, 234, 241
services, iv, vii, viii, ix, 2, 3, 4, 5, 6, 7, 8, 9, 11, 12, 13, 14, 15, 16, 17, 18, 19, 20, 21, 22, 23, 24, 25, 26, 27, 28, 29, 30, 31, 32, 33, 34, 35, 36, 37, 38, 39, 40, 41, 42, 43, 44, 45, 46, 47, 48, 49, 50, 51, 52, 53, 54, 55, 56, 57, 58, 59, 60, 61, 62, 63,64, 66, 67, 68, 69, 71, 72, 73, 74, 76, 78, 80, 101, 102, 104, 106, 108, 109, 110, 111, 115, 124, 146, 154, 155, 156, 158, 159, 165, 170, 173, 174, 175, 178, 179, 180, 181, 186, 187, 191, 193, 194, 195, 196, 198, 205, 208, 215, 238
shared computing resources, vii, viii, 2
Social Security, 2, 5, 6, 24, 27, 31, 32, 36, 44, 51, 61, 66, 67, 77, 79, 101, 141
Social Security Administration, 2, 5, 6, 24, 27, 31, 32, 36, 44, 51, 61, 66, 67, 77, 79, 101, 141
solution, ix, 11, 45, 63, 113, 114, 116, 124, 147, 148, 153, 155, 156, 158, 173, 183, 186, 187, 190, 193, 232
spending, vii, viii, 2, 4, 5, 7, 15, 16, 18, 29, 30, 31, 32, 33, 34, 38, 40, 53, 57, 66, 69, 70, 71, 78, 79, 80
state, 4, 14, 61, 65, 115, 118, 148, 206, 207, 209, 211, 213, 217, 219, 220, 224, 240, 242
storage, 5, 12, 13, 42, 49, 115, 116, 118, 156, 168, 169, 172, 177, 179, 181, 185, 189, 205, 207, 233, 240
Supreme Court, v, x, xii, xiii, 200, 202, 210, 212, 213, 237, 238, 239, 242

T

talent, 75, 160, 162, 163, 164, 171
taxonomy, 16
taxpayers, 159
technical assistance, 104
technical comments, 9, 61, 73
techniques, 149, 159, 162, 173
technological progress, 151
territorial, x, xi, 199, 200, 201, 210, 215, 234, 239
terrorism, 203, 226, 230
terrorist attack, 201
testing, 46, 112, 114, 115, 170
time frame, 22, 23, 24, 35, 38, 55, 113
training, 46, 48, 53, 122, 148, 149, 155, 160, 161, 164, 176

transparency, 13, 47, 52, 158, 173, 176, 180, 189, 225
Treasury, 2, 5, 6, 18, 25, 37, 39, 41, 44, 47, 50, 51, 52, 60, 64, 66, 67, 71, 102, 103
treaties, x, xii, 200, 202, 217, 218, 219, 220, 222, 223, 241

U

United Kingdom, 210, 229
United Nations, 220
United States, v, viii, x, xi, xii, xiii, 1, 16, 46, 59, 84, 97, 111, 145, 164, 168, 199, 200, 201, 202, 203, 204, 206, 207, 208, 209, 210, 211, 212, 213, 214, 215, 216, 217, 218, 219, 220, 221, 222, 223, 224, 227, 229, 233, 234, 235, 236, 237, 238, 239, 240, 241
United States v. Microsoft Corp, viii, xii, xiii, 201, 202, 204, 208, 209, 210, 212, 237

W

warfighter, vii, ix, 167, 169, 171, 172, 177, 180, 185, 187
warfighting, vii, 87, 172, 179, 181
web, 48, 106, 109, 110, 119, 120, 207, 239
web browser, 110
websites, 44, 47, 109, 110, 119, 120
workforce, vii, ix, 51, 53, 146, 149, 151, 155, 159, 160, 161, 163, 164, 165, 169, 175, 176, 182, 185, 189, 191
working groups, 155
worldwide, 48, 194, 232, 234